AF360917

Design and Applications of Additive Manufacturing and 3D Printing

Design and Applications of Additive Manufacturing and 3D Printing

Editor

Mika Salmi

MDPI • Basel • Beijing • Wuhan • Barcelona • Belgrade • Manchester • Tokyo • Cluj • Tianjin

Editor
Mika Salmi
Aalto University
Finland

Editorial Office
MDPI
St. Alban-Anlage 66
4052 Basel, Switzerland

This is a reprint of articles from the Special Issue published online in the open access journal *Designs* (ISSN 2411-9660) (available at: https://www.mdpi.com/journal/designs/special_issues/am_3d).

For citation purposes, cite each article independently as indicated on the article page online and as indicated below:

LastName, A.A.; LastName, B.B.; LastName, C.C. Article Title. *Journal Name* **Year**, *Volume Number*, Page Range.

ISBN 978-3-0365-3252-3 (Hbk)
ISBN 978-3-0365-3253-0 (PDF)

Contents

About the Editor

Mika Salmi works as a Research Director at the Aalto University Digital Design Laboratory and as a Staff Scientist at the Department of Mechanical Engineering. He completed his Ph.D. in 2013, focusing on medical applications of additive manufacturing in surgery and dental care. Salmi has published more than 60 scientific and technical papers on industrial and medical additive manufacturing applications. In his research, he takes advantage of whole-AM workflows and recent developments in materials, simulation, AM design, post-processing, and their applications. He is currently involved in projects related to the 3D printing of spare parts and medical applications, 4D printing, and researching industrial additive manufacturing applications.

Preface to "Design and Applications of Additive Manufacturing and 3D Printing"

Additive manufacturing (AM), more commonly known as 3D printing, has grown trememdously in recent years. It has shown its potential uses in the medical, automotive, aerospace, and spare part sectors. Personal manufacturing, complex and optimized parts, short series manufacturing, and local on-demand manufacturing are just some of its current benefits. The development of new materials and equipment has opened up new application possibilities, and equipment is quicker and cheaper to use when utilizing the new materials launched by vendors and material developers. AM has become more critical for the industry but also for academics. Since AM offers more design freedom than any other manufacturing process, it provides designers with the challenge of designing better and more efficient products.

The objective of this Special Issue was to provide a forum for researchers and practitioners to exchange their latest achievements and to identify critical issues and challenges for future investigations of the design and applications of additive manufacturing. The Special Issue consists of five original full-length articles on the topic.

Mika Salmi
Editor

Editorial

Design and Applications of Additive Manufacturing and 3D Printing

Mika Salmi

Department of Mechanical Engineering, Aalto University, 02150 Espoo, Finland; mika.salmi@aalto.fi

Citation: Salmi, M. Design and Applications of Additive Manufacturing and 3D Printing. *Designs* **2022**, *6*, 6. https://doi.org/10.3390/designs6010006

Received: 14 January 2022
Accepted: 15 January 2022
Published: 19 January 2022

Publisher's Note: MDPI stays neutral with regard to jurisdictional claims in published maps and institutional affiliations.

1. Introduction

Additive manufacturing (AM), or commonly, 3D printing, has been witnessed in various applications and purposes such as industrial applications in consumer products, energy, aerospace, medical, spare parts [1–6] and research and development activities such as prototyping, testing and material development [7–9]. What is common to all of these is that AM is not possible without the 3D model resulting from the design phase. Since AM allows the manufacture of complex geometries, we can improve the performance of the products through that. However, it is challenging to specify the best geometry for each purpose since we can optimize different things such as the mass, strength, price, manufacturing speed or performance. In addition, different AM processes require different approaches. This Special Issue offers a platform for researchers to study and report different design aspects of AM by taking into account differences between AM processes. Five original articles were published in this issue.

For the powder bed fusion of Ti6Al4V, Akram et al. [10] established a linear relationship between percentage elongation and the combined size of α lath and powder layer thickness using the rule of mixtures. They studied the microstructure's effect on the magnitude and anisotropy of the resultant mechanical properties, such as the yield strength and elongation. A Hall–Petch relationship was established between the α lath size and the yield strength magnitude for the as-built, heat-treated, transverse, and longitudinal built samples. Percentage elongation was affected by both α lath size and powder layer thickness, due to its correlation with the prior β columnar grain size.

To increase the spanwise length of the wing of unmanned aerial vehicles (UAV), Bishay et al. [11] studied a new design for the core of span-morphing. The purpose of morphing the wingspan is to increase lift and fuel efficiency. The main components that make up the structure are zero Poisson's ratio honeycomb substructure, telescoping carbon fiber spars and a linear actuator. The design maintains the airfoil shape and cross-sectional area during morphing because of its transverse rigidity and spanwise compliance. The wing model was analyzed computationally, manufactured, assembled and experimentally tested.

Guo et al. [12] compared the shape, porosity and mean curvature between triply periodic minimal surfaces and tetrahedral implicit surfaces. For replacing the conventional Cartesian coordinates, they developed a new coordinates system based on the perpendicular distances between a point and the tetrahedral faces to capture the periodicity of an implicit tetrahedral surface. For the triply periodic minimal surface, various tetrahedral implicit surfaces, including P-, D- and G-surfaces, were defined by combinations of trigonometric functions. For demonstration, they made a femur scaffold construction to demonstrate the process of modeling porous architectures using the implicit tetrahedral surface.

Topology optimization and laser-based powder bed fusion additive manufacturing were utilized by Orne et al. [13] to develop space flight hardware. They redesigned and manufactured two heritage parts for a Surrey Satellite Technology LTD (SSTL) Technology Demonstrator Space Mission and a system of five components for SpaceIL's lunar launch vehicle. During topology optimization, they incorporated AM manufacturing constraints such as the minimization of support structures, removing the unsintered powder and the minimization of heat transfer jumps. The designs were verified by finite element analysis. The components were fabricated, and the AM artifacts and in-process testing coupons have undergone verification and qualification testing.

Chadha et al. [14] explored the AM-enabled combination of function approach for the design of modular products. The approach replaced sub-assemblies within a modular product or system with more complex consolidated parts. This can increase the reliability of systems and products by reducing the number of interfaces and optimizing the more complex parts during the design. The fewer parts and the ability of users to replace or upgrade the system or product parts on-demand should reduce user risk and life cycle costs and prevent obsolescence. A case study for redesigning the AIM-120 AMRAAM Airframe was presented to illustrate the concept.

These studies show many different design and optimization paths for AM. Part consolidation, topology optimization, porous and lattice structures, modularity and optimizing mechanical properties can be performed with AM through design and AM process parameters. This opens a window where possibilities are endless, but finding the best ones is highly challenging. Moreover, the development of new AM processes and materials such as 4D printing will produce totally new challenges for design since dynamic behavior after the 3D printing needs to be considered. These could example include combining topology optimization with 4D printing and functionally graded adaptive metamaterials [15,16].

Funding: JAES Foundation.

Institutional Review Board Statement: Not applicable.

Informed Consent Statement: Not applicable.

Data Availability Statement: Not applicable.

Acknowledgments: All the articles were refereed through a double-blind peer-review process. The Guest Editor would like to thank all the authors for their excellent contributions. Special thanks to the reviewers for their valuable and critical comments significantly improved the articles. Thanks to the staff of the *Designs* journal, in particular, Ryan Pei, for his efforts and kind assistance.

Conflicts of Interest: The author declares no conflict of interest.

References

1. Bogers, M.; Hadar, R.; Bilberg, A. Additive manufacturing for consumer-centric business models: Implications for supply chains in consumer goods manufacturing. *Technol. Forecast. Soc. Change* **2016**, *102*, 225–239. [CrossRef]
2. Sun, C.; Wang, Y.; McMurtrey, M.D.; Jerred, N.D.; Liou, F.; Li, J. Additive manufacturing for energy: A review. *Appl. Energy* **2021**, *282*, 116041. [CrossRef]
3. Kestilä, A.; Nordling, K.; Miikkulainen, V.; Kaipio, M.; Tikka, T.; Salmi, M.; Auer, A.; Leskelä, M.; Ritala, M. Towards space-grade 3D-printed, ALD-coated small satellite propulsion components for fluidics. *Addit. Manuf.* **2018**, *22*, 31–37. [CrossRef]
4. Mäkitie, A.; Paloheimo, K.S.; Björkstrand, R.; Salmi, M.; Kontio, R.; Salo, J.; Yan, Y.; Paloheimo, M.; Tuomi, J. Medical applications of rapid prototyping–three-dimensional bodies for planning and implementation of treatment and for tissue replacement. *Duodecim* **2010**, *126*, 143–151. [PubMed]
5. Chekurov, S.; Salmi, M. Additive manufacturing in offsite repair of consumer electronics. *Phys. Procedia* **2017**, *89*, 23–30. [CrossRef]
6. Najmon, J.C.; Raeisi, S.; Tovar, A. Review of additive manufacturing technologies and applications in the aerospace industry. *Addit. Manuf. Aerosp. Ind.* **2019**, *11*, 7–31.
7. Pham, D.T.; Gault, R.S. A comparison of rapid prototyping technologies. *Int. J. Mach. Tools Manuf.* **1998**, *38*, 1257–1287. [CrossRef]
8. Scaravetti, D.; Dubois, P.; Duchamp, R. Qualification of rapid prototyping tools: Proposition of a procedure and a test part. *Int. J. Adv. Manuf. Technol.* **2008**, *38*, 683–690. [CrossRef]
9. Nilsén, F.; Ituarte, I.F.; Salmi, M.; Partanen, J.; Hannula, S.P. Effect of process parameters on non-modulated Ni-Mn-Ga alloy manufactured using powder bed fusion. *Addit. Manuf.* **2019**, *28*, 464–474. [CrossRef]
10. Akram, J.; Pal, D.; Stucker, B. Establishing Flow Stress and Elongation Relationships as a Function of Microstructural Features of Ti6Al4V Alloy Processed using SLM. *Designs* **2019**, *3*, 21. [CrossRef]
11. Bishay, P.L.; Burg, E.; Akinwunmi, A.; Phan, R.; Sepulveda, K. Development of a New Span-Morphing Wing Core Design. *Designs* **2019**, *3*, 12. [CrossRef]
12. Chadha, C.; Crowe, K.A.; Carmen, C.L.; Patterson, A.E. Exploring an AM-Enabled Combination-of-Functions Approach for Modular Product Design. *Designs* **2018**, *2*, 37. [CrossRef]
13. Guo, Y.; Liu, K.; Yu, Z. Tetrahedron-Based Porous Scaffold Design for 3D Printing. *Designs* **2019**, *3*, 16. [CrossRef]

14. Orme, M.; Madera, I.; Gschweitl, M.; Ferrari, M. Topology Optimization for Additive Manufacturing as an Enabler for Light Weight Flight Hardware. *Designs* **2018**, *2*, 51. [CrossRef]
15. Bodaghi, M.; Damanpack, A.; Liao, W.H. Adaptive metamaterials by functionally graded 4D printing. *Mater. Des.* **2017**, *135*, 26–36. [CrossRef]
16. Zolfagharian, A.; Denk, M.; Bodaghi, M.; Kouzani, A.Z.; Kaynak, A. Topology-optimized 4D printing of a soft actuator. *Acta Mech. Solida Sin.* **2019**, *13*, 137. [CrossRef]

Article

Establishing Flow Stress and Elongation Relationships as a Function of Microstructural Features of Ti6Al4V Alloy Processed Using SLM

Javed Akram *, Deepankar Pal and Brent Stucker

ANSYS Inc., 1794 Olympic Parkway # 110, Park City, UT 84098, USA; Deepankar.pal@ansys.com (D.P.); brent.stucker@ansys.com (B.S.)
* Correspondence: Javed.akram@ansys.com

Received: 14 February 2019; Accepted: 11 April 2019; Published: 13 April 2019

Abstract: Selective laser melting (SLM) is an attractive technology for fabricating complex metal parts with reduced number of processing steps compared to traditional manufacturing technologies. The main challenge in its adoption is the variability in mechanical property produced through this process. Control and understanding of microstructural features affected by the SLM process is the key for achieving desirable mechanical properties. Numerous studies have been published related to microstructure and mechanical properties of SLM printed parts; however, few of those reported end-to-end process–structure–property relationship. Therefore, the current study aims to comprehensively present the widespread microstructure information available on SLM processed Ti6Al4V alloy. Furthermore, its effects on the magnitude and anisotropy of the resultant mechanical properties, such as the yield strength and elongation, has been established. A Hall–Petch relationship is established between α lath size and yield strength magnitude for the as-built, heat-treated, transverse, and longitudinal built samples. The anisotropy in flow stress is established using the α lath size and prior β grain orientation. Percentage elongation was identified to be affected by both α lath size and powder layer thickness, due to its correlation with the prior β columnar grain size. A linear relationship was established between percentage elongation and combined size of α lath and powder layer thickness using the rule of mixtures.

Keywords: selective laser melting (SLM); Ti6Al4V; structure–property relationship; microstructure; Hall–Petch relationship

---

1. Introduction

Additive manufacturing (AM) is a method of fabricating 3D components where materials are added in a layer-by-layer fashion [1]. It is an advanced manufacturing technology capable of fabricating complex part geometries with low lead time and material consumption. Metal-melting AM technologies, their feed systems, and energy sources are described elsewhere [2]. Among these, selective laser melting (SLM) and electron beam melting (EBM) are widely used methods in order to avail printing of complex geometries. A big challenge in the adoption of these methods is the variability in mechanical properties, which can vary from machine to machine as well as with changing process parameters. There are many process parameters which are responsible for this variability such as the power, velocity and type of heat source (e.g., electron beam or laser, laser scan strategy, powder layer thickness, hatch spacing, beam diameter, pre-heating) [3]. These parameters vary considerably with machine type and most of them are either confidential or locked for a machine type and a material system. As a result, it can be very challenging to reproduce a part with the same mechanical properties using different machines. This leads to the motivation behind exploration for a unified approach to correlate process parameter variations with resultant mechanical properties of additively manufactured materials, using a thorough understanding of microstructural features and morphologies. The current study aims to look into the published microstructure information and combine results in a meaningful form to demonstrate how microstructure relates to the mechanical properties of a part processed through SLM. The material system investigated in this article is Ti6Al4V due to its applications in a variety of industries, such as automotive, aerospace, and medical sectors.

2. Background

Ti6Al4V is an attractive alloy for lightweight structural components due to its excellent strength-to-weight ratio, corrosion resistance, and bio combability [4]. This alloy contains 6 wt.% of aluminum and 4 wt.% of vanadium, which are α and β stabilizers, respectively. The α stabilizers (Al, O_2, N_2, C, etc.,) extend the α phase field to higher temperatures, whereas β stabilizers (V, Mo, Ta, etc.,) shift the β phase field to lower temperatures. Depending on the starting temperature, cooling rate, thermomechanical process, and post processing heat treatment, the final microstructures could be lamellar, martensitic, or equiaxed [4,5]. For example, when heated fully above the β-transus temperature of 995 °C followed by a furnace quasi-static cooling, the final microstructure consists of a lamellar structure of α and β, as shown in Figure 1a. Depending on the cooling rate, the resulting width of the lamellae could be coarsened or made finer. Rapid cooling leads to a martensitic transformation as shown in Figure 1b, which offers high strength-to-weight ratio, caused by high dislocation densities as a function of the severity of cooling. Equiaxed grains, as shown in Figure 1c, formed during recrystallization heat treatment on pre-deformed microstructures, also increase strength. Depending upon the heat treatment temperature, the size of equiaxed grains could be large or small. Further, at different combinations of solution heat treatments, the final distribution of microstructure could be bimodal in some cases, such as a mix of equiaxed and lamellar structures [4,6,7], as shown in Figure 1d. These microstructures have their specific roles on the mechanical behavior. Based on the literature [4,6], it could be established that the fine microstructural features, such as the grain size, offers high strength and ductility; coarse microstructure offers better resistance to creep and fatigue crack growth; equiaxed microstructure offers good ductility and fatigue strength; and lamellar microstructure offers better fracture toughness as well as superior resistance to creep and fatigue crack growth.

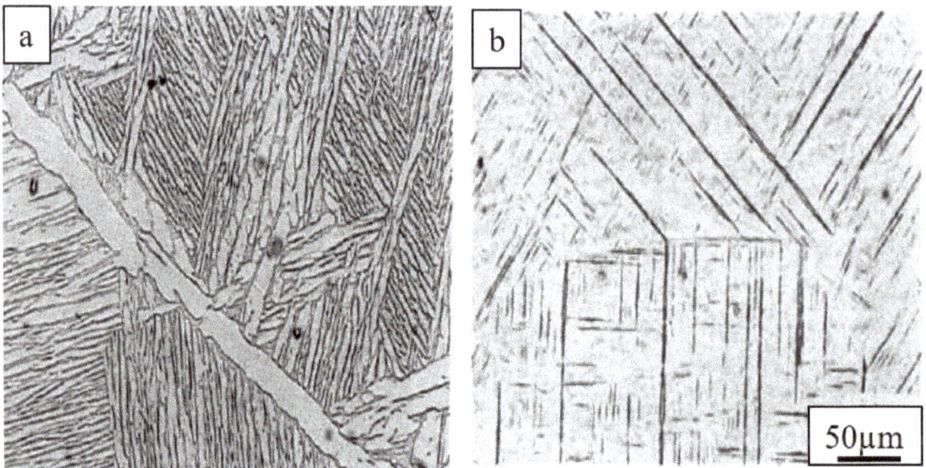

Figure 1. *Cont.*

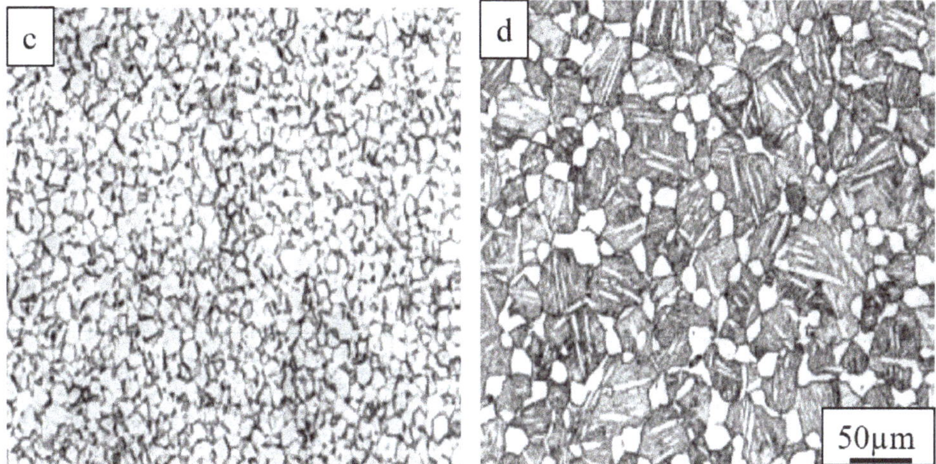

Figure 1. Example of (**a**) lamellar, (**b**) acicular, (**c**) equiaxed, and (**d**) bi-modal microstructure in Ti6Al4V alloy [4,6].

3. Results and Discussion

In SLM processed materials, the as-built microstructure is mainly composed of a very fine α' lath martensite structure due to excessively high rates of cooling. The typical α' lath martensite micro structure as processed using the SLM method is shown in Figure 2. Table 1 summarizes the literature review of all the as-built and heat-treated microstructures (i.e., α colony width and prior β columnar grain width), processed using different process parameters and corresponding mechanical property variations. The resolution of fine α' lath martensite ranged from 0.2 to 0.6 μm. In few cases, these features could grow up to 2–3 μm during the heat treatment. Presence of acicular α' in as-built structures imparts a very high amount of dislocation density [8], which is responsible for high strength compared to conventional processed materials. However, this high strength is compromised by reduced elongation. Therefore, a post process heat treatment becomes essential. It can be seen from the literature data presented in Table 1 that after post additive heat treatment (PAHT), a considerable amount of gain in elongation value is achieved compared to the as-built parts at the cost of yield strength, which is comparable to that of the wrought products.

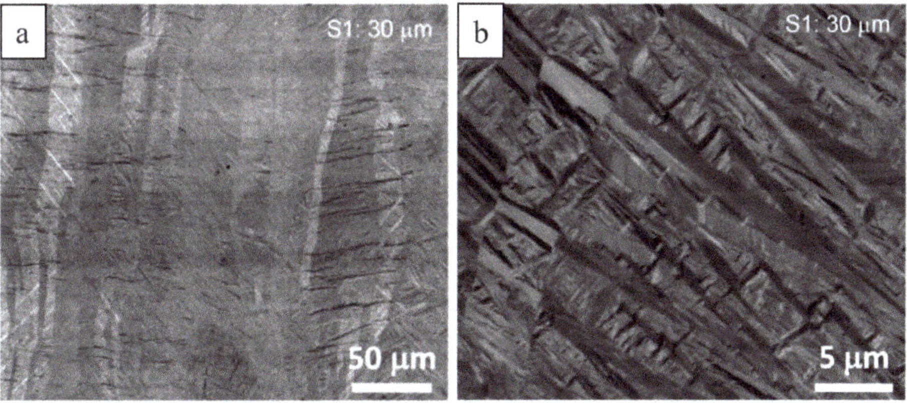

Figure 2. Typical α' lath martensite micro structure as processed using SLM method: (**a**) Lower and (**b**) higher magnification [9].

Table 1. List of microstructures (α lath and prior β columnar width) and corresponding mechanical property from literature. The mechanical data are included from the work which reported mechanical property along with their microstructure information.

Reference	Process Parameter *P(W), V(mm/s), HS (μm), LT (μm), LHI (J/mm³), SS, SP, PH (°C)	Sample Orientation	Microstructure W is Width in (μm)	YS (MPa)	UTS (MPa)	EL (%)	E (GPa)
[10]	P: 200, V: 200, HS: 180, LT: 50 μm, LHI: 111, SP: bidirectional scan vector with 67° rotated, MTT SLM 250	L	Acicular α′, β columnar grains (W: 109.48)	910 ± 9.9	1035 ± 29	3.3 ± 0.76	-
[11]	P: 160, V: 600, HS: 200, LT: 40, LHI: 33, BD: 220, SP: bidirectional scan vector with 90° rotated, PH: 500, Trumpf LF 250	L	Acicular α′ (W: 0.36)	1137 ± 20	1206 ± 8	7.6 ± 2	105 ± 5
		T		962 ± 47	1166 ± 25	1.7±0.3	102 ± 7
	Heat treated at 950 °C for 1h followed by water quenching	L	Acicular α′ + α + β (W: 1.7)	944 ± 8	1036 ± 30	8.5 ± 1	103 ± 11
		T		925 ± 14	1040 ± 4	7.5 ± 2	98 ± 3
[12]	P: 175, V: 710, HS: 120, LT: 30 μm, LHI: 68.5, SP: bidirectional scan vector with 79° rotation, MTT SLM250	L	Acicular α′, β columnar grains (W: 117.2)	1166 ± 6	1321 ± 6	2.0 ± 0.7	-
	P: 250, V: 1600, HS: 60, LT: 30, LHI: 86.8, SS: 50, SP: bidirectional scan vector with 90° rotation	L	Acicular α′, β columnar grains (W: 55.5)	1110 ± 9	1267 ± 5	7.28 ± 1.12	109.2 ± 3.1
[7]	Heat treated at 850 °C for 2 h, followed by furnace cooling	L	Mixture of α + β (W:1.27 ± 0.13, V: 73%), β columnar grains (W: 82.17)	955 ± 6	1004 ± 6	12.84 ± 1.36	114.7 ± 3.6
	Heat treated at 940 °C for 1 h, followed by 650 °C for 2 h, then air cooled	L	Lamellar mixture of α + β (W: ~2), β columnar grains (W: 82.17)	899 ± 27	948 ± 27	13.59 ± 0.32	115.5 ± 2.4
[13]	P: 157, V: 225, HS: 100, LT: 50, LHI: 139.5, SS: 70, SP: bidirectional scan vector with 67° rotated, flat sample, Renishaw AM250	L, XY plane	Acicular α′ (W: 0.57 ± 0.13, L: 8 ± 3), β columnar grains (W: XY: 91.29, XZ: 89.61, and YZ: 76.68)	1075 ± 25	1199 ± 49	7.6 ± 0.5	113 ± 5
	Flat sample	L, XZ plane		978 ± 5	1143 ± 6	11.8 ± 0.5	115 ± 6
		T, ZX plane		967 ± 10	1117 ± 3	8.9 ± 0.4	119 ± 7
	Stress relieved at 730 °C for 2 h, FC at 283.15 K/min	L, XY plane	Mixture of α + β (W: 1.2 ± 0.3, L: 8.7 ± 2.4), β columnar grains (W: XY: 91.29, XZ: 89.61, and YZ: 76.68)	974 ± 7	1065 ± 21	7.0 ± 0.5	112 ± 6
		L, XZ plane		958 ± 6	1057 ± 8	12.4 ± 0.7	113 ± 9
		T, ZX plane		937 ± 9	1052 ± 11	9.6 ± 0.9	117 ± 6
[14,15]	P: 194, V: 1000, HS: 80, LT: 20, LHI: 121.25, SP: bidirectional scan vector with 90° rotation	L	Acicular α′ (W: ~1.5), β columnar grains (W: 53.7)	937.95	1140.8	4.2	-
		T		853.5	1077.5	4.5	-

Table 1. *Cont.*

Reference	Process Parameter *P(W), V(mm/s), HS (μm), LT (μm), LHI (J/mm³), SS, SP, PH (°C)	Sample Orientation	Microstructure W is Width in (μm)	YS (MPa)	UTS (MPa)	EL (%)	E (GPa)
[16]	P: 120–200, BD: 200–600	L	Acicular α′	990 ± 5	1065 ± 10	8.1 ± 0.3	-
	Heat treatment variant 1: NI	L	Lamellar α + β	835 ± 5	915 ± 10	10.6 ± 0.6	
	Heat treatment variant 2: NI	L	Lamellar α + β and Globular α	870 ± 15	990 ± 15	11.0 ± 0.5	
[17]	P: 200, V: 1250, HS: ~100, LT: 40, LHI: 40, SS: 250, SP: bidirectional scan vector with 90° rotation, Concept Laser M2	T	Acicular α′ (W: <0.5)	986	1155	10.9	112.4
	Heat treated at 700 °C for 1 h, 10 K/min cool	T	Acicular α′ (W: <1)	1051	1115	11.3	117.4
	Heat treated at 900 °C for 2 hours + 700 °C for 1 h, 10 K/min	T	Lamellar α + β (W: 2–3, L: 50–60)	908	988	9.5	118.8
	HIP at 900 °C and 100 MPa for 2 h in Ar₂ gas atmosphere + 700 °C for 1 h, 10 K/min cool.	T	Lamellar α + β (W: 2–3, L: 50–60)	885	973	19.0	115.4
[18]	P: 175, V: 710, HS: 120, LT: 30, PH: 100, LHI: 68.46, SLM 250^HL	T	Acicular α′ (W: 0.37 ± 0.07)	1008 ± 30	1080 ± 30	1.6 ± 2	-
	Heat treated at 800 °C for 2 h in Ar₂ gas, FC	T	Mixture of α + β (W: 0.57 ± 0.06)	962 ± 30	1040 ± 30	5 ± 2	-
	HIPed at 920 °C and 1000 bar for 2 h in Ar₂ gas, FC	T	Lamellar α + β (W: 2.38 ± 0.3) and Globular α (S: 5.0 ± 1.6)	912 ± 30	1005 ± 30	8.3 ± 2	-
	Heat treated at 1050 °C for 2 h in vacuum, FC	T	Lamellar α+β (W: 9.75 ± 3.7) and If globular (S: 13.73 ± 5.3)	798 ± 30	945 ± 30	11.6 ± 2	-
[19]	No information	L	Acicular α′ (W: 0.1 to 0.3)	1330	1400	4.4	-
[20]	P: 5500, V: 10⁴, LT: 40, LHI: 0.55, PH: 700, EOSINT M270	T	Acicular α′ (W: 0.23 to 0.3)	~850	~940	6.5	-
[21]	P: 375, V: 1029, HS: 120, LT: 60, FOD: 2, T_i: 1, LHI: 50.62, SP: bidirectional scan vector with 90° rotation, SLM 250^HL	T	Lamellar α + β (W: 0.52 ± 0.22)	1022 ± 10	1090 ± 10	12.7 ± 2.1	
	T_i: 5 and all the parameters remain same	T	Lamellar α + β (W: 0.29 ± 0.13)	1093 ± 15	1149 ± 11	11.3 ± 0.5	
	T_i: 8 and all the parameters remain same	T	Lamellar α + β (W: 0.25 ± 0.10)	1112 ± 3	1165 ± 2	11.6 ± 1.2	

*L: Longitudinal (long length along x or y direction); T: Transverse (long length along build direction); P: Power; V: Velocity; HS: Hatch Spacing; LT: Layer Thickness; LHI: Laser Heat Intensity; SS: Spot Size; SP: Scan Pattern; PH: Platform Heating; YS: Yield Strength; UTS: Ultimate Tensile Strength; EL: Elongation; E: Modulus of Elasticity; FOD: Focal Offset Distance; T_i: Inter Layer Time; HIP: Hot Isostatic Pressure; in longitudinal samples, columnar grains are transverse to loading direction; in transverse samples, columnar grains are along the loading direction.

3.1. Correlation between As-Built and Heat-Treated Yield Strength and α/β Feature Sizes

The presented information in Table 1 clearly indicates that the width of α lath size is mainly responsible for high yield strengths. Figure 3 illustrates the Hall–Petch relationship between α lath size and yield strength of as-built and heat-treated samples. Although the inversely proportional relationship between grain size and flow stress is known from Hall–Petch expression, the constant of proportionality needs to be determined as a function of processing condition and given material. Using information from literature as presented in Table 1, such constants have been evaluated for Ti6Al4V alloy processed through SLM, as shown in Figure 3. The α lath size values presented in Table 1 were either taken as reported in a paper or calculated using image J software. It clearly reflects the fact that the finer the lath size is, the higher the resulting yield strength is, and is in good agreement with the conventional Hall–Petch relationship. It could be also observed that the samples built in the longitudinal direction show higher strength when compared with their transverse counterparts, which is common in all AM parts [22,23]. Although, the strength increases faster with respect to inverse square root α lath size in the longitudinal direction than its transverse counterpart. This anisotropy in mechanical behavior is not only observed in SLM but also in EBM- and DED-produced parts, and could be attributed to strong texturing during SLM processing [24]. Texture in the AM parts occurs from growth of columnar grains, which seem to preferentially occur in an epitaxial manner from one layer to multiple layers in the building direction [24], driven due to strong −z thermal gradients towards the heat sink. In Ti6Al4V, the columnar grains are a result of prior β grains, which are stable above the β-transus temperature. Upon cooling, α grains nucleate at the columnar β grain boundaries and grow according to the Burger orientation correlation during cooling [25]. Due to the rapid cooling in SLM, these α grains grow in a very fine needle-like shape (acicular α′) with certain variant selection [25] by diffusionless transformation mode. Therefore, the β phase is almost absent in the final microstructure. These prior β grains in the as-built microstructure cannot be seen but are important in evaluating final yield strength and elongation of a material, because final orientation of α laths are dependent on the orientation of the prior β columnar grains. When these columnar grains are oriented transverse to the loading axis, they impart high yield strength by dislocation pile-up at prior β columnar grain boundaries. In longitudinally/horizontally built samples, these columnar grains are oriented transverse to the loading direction and; therefore, high yield strength values are seen compared to transverse, as shown in Figure 4. The grain boundaries of these columnar grains are not easy to identify and are sometimes misinterpreted by laser vector hatch spacing. Careful identification or reconstruction of β grains using electron backscattered diffraction data is a promising technique to resolve the width of these grains [26].

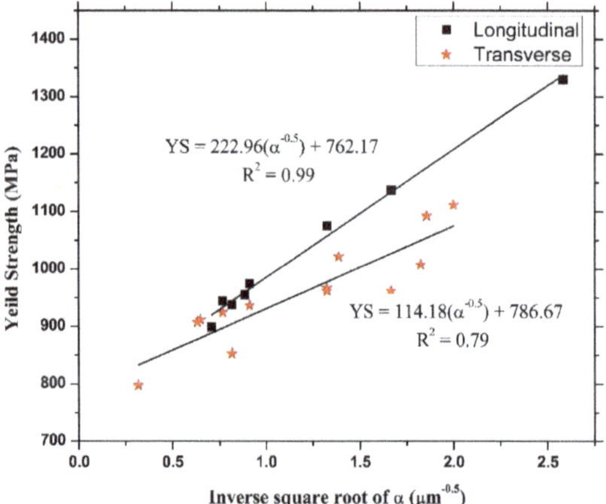

Figure 3. Hall–Petch relationship between α lath size and yield strength for as-built and heat-treated samples.

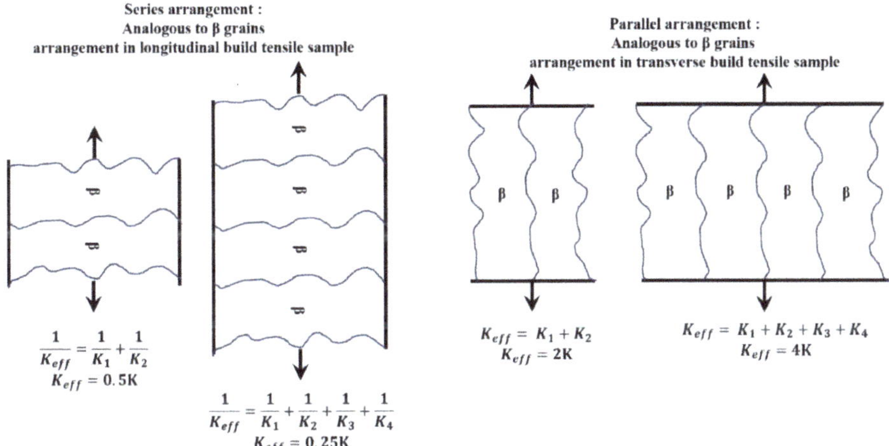

Figure 4. Schematic of spring arrangement in series and parallel case analogous to β grains in longitudinal and transverse built samples.

3.2. Correlation between As-Built Elongation and α/β Feature Sizes

Apart from the yield strength, another important property specification of any engineering material required for the structural integrity is ductility. Figure 5a is the plot between α lath size and percent elongation of as-built and heat-treated SLM samples. It can be seen from the plot that there is no clear relationship between elongation and α lath size, as seen for yield strength, as well as no distinction in elongation values between longitudinally and transversely built samples. The wide scatter indicates α lath size is not the only factor responsible for percent elongation. It could be possible that prior β grain size and orientation does affect elongation values in a similar manner as yield strength values. Although, the width of β columnar grains are not widely reported in the literature since β phase becomes unstable at lower temperatures. By incorporating larger amounts of β stabilizers, the β microstructure could be frozen at a relatively lower temperature to study these effects, if any. However, powder layer thickness does appear to relate to elongation. Higher powder layer thickness combined with large α lath size imparts high percentage elongation. Rule of mixtures is adopted to combine these two parameters, where 0.8 and 0.2 fraction of contribution is given to α lath size and powder layer thickness, respectively (0.8*(α lath size) + (0.2*layer thickness)). The combined size is plotted with percentage elongation for as-built samples, as shown in Figure 5b. It can be observed from the plot that a linear relationship is established, which shows layer thickness does affect the elongation values.

The powder layer thickness could be corelated with prior β columnar grain size by considering the nucleation population in a specified build height. For example, the higher the layer thickness, the lesser would be the number of melted layers for a specified build height. This considerably reduces the total number of nucleation events, hence a larger prior β columnar grain width becomes probable. This is also indicated in Table 1 with a few exceptions. By relating β grain size with powder layer thickness, it can be inferred that a larger prior β columnar grain width combined with a large α lath width imparts higher elongation as shown in Figure 5b. Like the yield strength relationship, the orientation of prior β columnar grain also affects elongation. Higher elongation values are achieved for transverse as-built samples when β columnar grains are aligned along the tensile loading axis, as shown in Figure 5b. This higher elongation is the result of a higher Schmid factor value, which is achieved due to nearly transverse orientation of α' (~60° with prior β columnar grains) laths with the tensile loading axis for transversely built samples [14]. Apart from higher elongation, the rate at which elongation increases as a function of combined α lath widths and powder layer thickness, is also faster in transverse samples compared to their longitudinal counterparts. One possible explanation is as follows: The effective spring coefficient decreases from 0.5 to 0.25 K resulting in a change of 0.25 K for loss of two springs (from four to two springs) in the series spring configuration on the left (Figure 4). This effectively implies that the stiffness change per unit spring is 0.125 K. Similarly, the stiffness change per unit spring is K for the parallel spring configuration. The reduction in the number of springs in the parallel and series scenario is theoretically equivalent to the increase in powder layer thickness. Increase in powder layer thickness corresponds to reduction in the number of nucleation spots, which leads to increase in grain size. The stiffness change per unit spring stiffness is higher in the transversely built samples when loaded in transverse direction compared to longitudinally built samples when loaded in longitudinal direction. Henceforth, the increase in elongation with respect to the decreasing number of grains or increasing powder layer thickness is also going to be higher in transverse samples compared to their longitudinal counterparts.

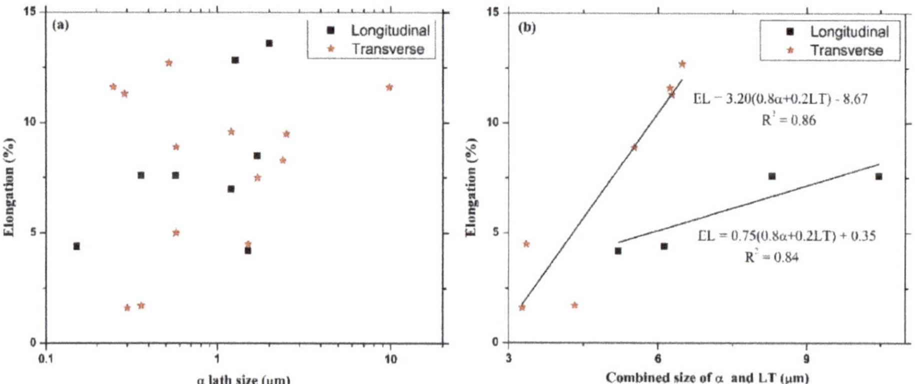

Figure 5. Plot between percentage elongation and (**a**) α lath size for as-built and heat-treated samples and (**b**) combined α lath and powder layer thickness (LT) using rule of mixture for as-built samples.

3.3. Correlation between Heat-Treated Elongation and α/β Feature Sizes

The percentage elongations of heat-treated samples for longitudinal and transverse oriented samples are plotted with combined α lath and layer thickness (relating with β columnar) width using rule of mixtures, as shown in Figure 6. The transverse-built heat-treated samples again showed good linear relationship with combined α lath and layer thickness; however, longitudinal samples showed an inverse relationship with increasing combined width. This discrepancy can be anticipated by considering the effect of PAHT conditions. In heat-treated samples; holding time, temperature, and cooling media are factors which affect the size of α lath and β columnar width. It can be seen from Table 1 that the lower elongation values correspond to the samples that were either heat treated at very low temperature or water quenched after heat treatment. Therefore, correct measurement of prior β columnar grains and its orientation in addition to α lath size is needed to correctly evaluate or to establish a relationship for yield strength and elongation.

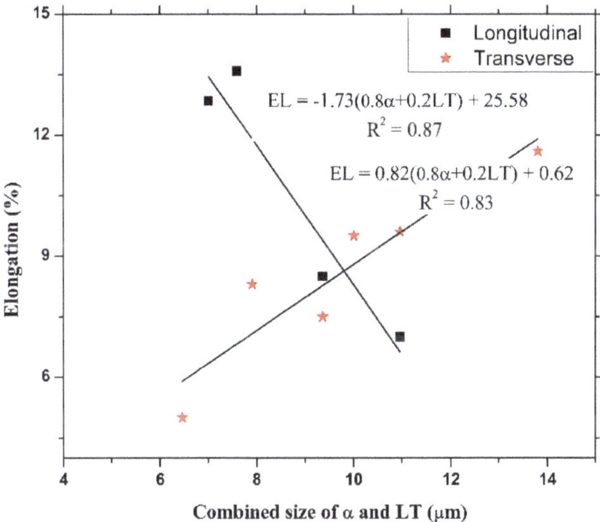

Figure 6. Plot between percentage elongation and combined α lath and powder layer thickness (LT) using rule of mixture for heat-treated samples.

4. Conclusions

It could be concluded that widely reported tensile data for Ti6Al4V fabricated through SLM, when combined with microstructural information (i.e., structure, α lath, and prior β columnar grain width size), enables a Hall–Petch relationship to be established, showing that finer α lath size is responsible for imparting high yield strength. Orientation of prior β columnar grains introduce anisotropy in mechanical behavior, which is commonly observed in AM-fabricated samples. Transverse orientation of prior β columnar grains with respect to tensile loading axis gives higher yield strength. Powder layer thickness showed considerable effect on elongation due to its correlation with β columnar grains. Rule of mixtures is adopted to establish a relationship between grain size and elongation. Results show good correlation between combined α lath, powder layer thickness (related to β columnar), and elongation.

Design Criteria: For a design perspective, the built material should have optimal elongation in the as-build condition along with strength. Based on the observed results and inferred trends, the combination of high strength and high elongation value can be achieved in the as-build condition by properly choosing the powder layer thickness. A high powder layer thickness, which is related to β columnar grains by the virtue of number density of nucleation sites, provides high elongation and conserves the high strength in a part. This work can be very useful to improve the mechanical properties of AM-printed parts by using larger layer thickness. As we mentioned, the powder layer thickness values are related with β columnar grains, in future it would be very interesting to see if this relationship still holds true using the β columnar grain size directly instead of layer thickness. If the value of α lath size and powder layer thickness are known, the yield strength and elongation values could be predicted or by using Integrated Computational Materials Engineering (ICME) tools which can predict these microstructure features.

Author Contributions: For research articles with several authors, a short paragraph specifying their individual contributions must be provided. conceptualization, J.A. and D.P.; writing—original draft preparation, J.A.; writing—review and editing, J.A., D.P., and B.S.

Funding: This research was funded by Naval Air Command, grant number N68335-17-C-0157.

Acknowledgments: The authors gratefully acknowledge the financial help rendered by the Naval Air Command (grant no. N68335-17-C-0157) and Mechanical Business Unit at ANSYS in supporting this work and its public dissemination.

Conflicts of Interest: The authors declare no conflict of interest.

References

1. Gibson, I.; Rosen, D.W.; Stucker, B. *Additive Manufacturing Technologies*; Springer: Berlin, Germany, 2010; Volume 238.
2. Frazier, W.E. Metal additive manufacturing: A review. *J. Mater. Eng. Perform.* **2014**, *23*, 1917–1928. [CrossRef]
3. Kumar, P.; Farah, J.; Akram, J.; Teng, C.; Ginn, J.; Misra, M. Influence of laser processing parameters on porosity in Inconel 718 during additive manufacturing. *Int. J. Adv. Manuf. Technol.* **2019**. [CrossRef]
4. Peters, M.; Hemptenmacher, J.; Kumpfert, J.; Leyens, C. Structure and Properties of Titanium and Titanium Alloys. In *Titanium and Titanium Alloys: Fundamentals and Applications*; Wiley: Hoboken, NJ, USA, 2003; pp. 1–36.
5. Ramosoeu, M.; Chikwanda, H.; Bolokang, A.; Booysen, G.; Ngonda, T. Additive manufacturing: Characterization of TI-6AI-4V alloy intended for biomedical application. In *Proceedings of the The Southern African Institute of Mining and Metallurgy Advanced Metals Initiative, Light Metals Conference*; Southern African Institute of Mining and Metallurgy: Muldersdrift, South Africa, 2010; pp. 337–344.
6. Lütjering, G.; Williams, J.C.; Gysler, A. Microstructure and mechanical properties of Titanium alloys. In *Microstructure and Properties of Materials*; World Scientific: Singapore, 2000; Volume 2, pp. 1–74.
7. Vrancken, B.; Thijs, L.; Kruth, J.-P.; van Humbeeck, J. Heat treatment of Ti6Al4V produced by Selective Laser Melting: Microstructure and mechanical properties. *J. Alloys Compd.* **2012**, *541*, 177–185. [CrossRef]
8. Cottam, R.; Palanisamy, S.; Avdeev, M.; Jarvis, T.; Henry, C.; Cuiuri, D.; Balogh, L.; Rashid, R.A.R. Diffraction Line Profile Analysis of 3D Wedge Samples of Ti-6Al-4V Fabricated Using Four Different Additive Manufacturing Processes. *Metals* **2019**, *9*, 60. [CrossRef]
9. Xu, W.; Brandt, M.; Sun, S.; Elambasseril, J.; Liu, Q.; Latham, K.; Xia, K.; Qian, M. Additive manufacturing of strong and ductile Ti-6Al-4V by selective laser melting via in situ martensite decomposition. *Acta Mater.* **2015**, *85*, 74–84. [CrossRef]
10. Edwards, P.; Ramulu, M. Fatigue performance evaluation of selective laser melted Ti-6Al-4V. *Mater. Sci. Eng. A* **2014**, *598*, 327–337. [CrossRef]
11. Vilaro, T.; Colin, C.; Bartout, J.D. As-fabricated and heat-treated microstructures of the Ti-6Al-4V alloy processed by selective laser melting. *Metall. Mater. Trans. A Phys. Metall. Mater. Sci.* **2011**, *42*, 3190–3199. [CrossRef]
12. Mertens, A.; Reginster, S.; Paydas, H.; Contrepois, Q.; Dormal, T.; Lemaire, O.; Lecomte-Beckers, J. Mechanical properties of alloy Ti-6Al-4V and of stainless steel 316L processed by selective laser melting: influence of out-of-equilibrium microstructures. *Powder Metall.* **2014**, *57*, 184–189. [CrossRef]
13. Simonelli, M.; Tse, Y.Y.; Tuck, C. Effect of the build orientation on the mechanical properties and fracture modes of SLM Ti-6Al-4V. *Mater. Sci. Eng. A* **2014**, *616*, 1–11. [CrossRef]
14. Yang, J.; Yu, H.; Wang, Z.; Zeng, X. Effect of crystallographic orientation on mechanical anisotropy of selective laser melted Ti-6Al-4V alloy. *Mater. Charact.* **2017**, *127*, 137–145. [CrossRef]
15. Yang, J.; Yu, H.; Yin, J.; Gao, M.; Wang, Z.; Zeng, X. Formation and control of martensite in Ti-6Al-4V alloy produced by selective laser melting. *Mater. Des.* **2016**, *108*, 308–318. [CrossRef]
16. Facchini, L.; Magalini, E.; Robotti, P.; Molinari, A.; Höges, S.; Wissenbach, K. Ductility of a Ti-6Al-4V alloy produced by selective laser melting of prealloyed powders. *Rapid Prototyp. J.* **2010**, *16*, 450–459. [CrossRef]
17. Kasperovich, G.; Hausmann, J. Improvement of fatigue resistance and ductility of TiAl6V4 processed by selective laser melting. *J. Mater. Process. Technol.* **2015**, *220*, 202–214. [CrossRef]
18. Leuders, S.; Thöne, M.; Riemer, A.; Niendorf, T.; Tröster, T.; Richard, H.A.; Maier, H.J. On the mechanical behaviour of titanium alloy TiAl6V4 manufactured by selective laser melting: Fatigue resistance and crack growth performance. *Int. J. Fatigue* **2013**, *48*, 300–307. [CrossRef]

19. Murr, L.E.; Quinones, S.A.; Gaytan, S.M.; Lopez, M.I.; Rodela, A.; Martinez, E.Y.; Hernandez, D.H.; Martinez, E.; Medina, F.; Wicker, R.B. Microstructure and mechanical behavior of Ti-6Al-4V produced by rapid-layer manufacturing, for biomedical applications. *J. Mech. Behav. Biomed. Mater.* **2009**, *2*, 20–32. [CrossRef]

20. Koike, M.; Greer, P.; Owen, K.; Lilly, G.; Murr, L.E.; Gaytan, S.M.; Martinez, E.; Okabe, T. Evaluation of titanium alloys fabricated using rapid prototyping technologies-electron beam melting and laser beam melting. *Materials* **2011**, *4*, 1776–1792. [CrossRef] [PubMed]

21. Xu, W.; Lui, E.W.; Pateras, A.; Qian, M.; Brandt, M. In situ tailoring microstructure in additively manufactured Ti-6Al-4V for superior mechanical performance. *Acta Mater.* **2017**, *125*, 390–400. [CrossRef]

22. Carroll, B.E.; Palmer, A.; Beese, A.M. Anisotropic tensile behavior of Ti-6Al-4V components fabricated with directed energy deposition additive manufacturing. *Acta Mater.* **2015**, *87*, 309–320. [CrossRef]

23. Beese, A.M.; Carroll, B.E. Review of mechanical properties of Ti-6Al-4V made by laser-based additive manufacturing using powder feedstock. *Jom* **2016**, *68*, 724–734. [CrossRef]

24. Akram, J.; Chalavadi, P.; Pal, D.; Stucker, B. Understanding grain evolution in additive manufacturing through modeling. *Addit. Manuf.* **2018**, *21*, 255–268. [CrossRef]

25. Simonelli, M.; Tse, Y.Y.; Tuck, C. On the texture formation of selective laser melted Ti-6Al-4V. *Metall. Mater. Trans. A Phys. Metall. Mater. Sci.* **2014**, *45*, 2863–2872. [CrossRef]

26. Antonysamy, A.A.; Meyer, J.; Prangnell, P.B. Effect of build geometry on the β-grain structure and texture in additive manufacture of Ti-6Al-4V by selective electron beam melting. *Mater. Charact.* **2013**, *84*, 153–168. [CrossRef]

Article

Development of a New Span-Morphing Wing Core Design

Peter L. Bishay *, Erich Burg, Akinwande Akinwunmi, Ryan Phan and Katrina Sepulveda

Department of Mechanical Engineering, California State University, Northridge, Northridge, CA 91330, USA;
erich.burg.303@my.csun.edu (E.B.); akinwande.akinwunmi.72@my.csun.edu (A.A.);
ryan.phan.25@my.csun.edu (R.P.); katrina.sepulveda.801@my.csun.edu (K.S.)
* Correspondence: peter.bishay@csun.edu; Tel.: +1-818-677-7803

Received: 4 January 2019; Accepted: 2 February 2019; Published: 7 February 2019

Abstract: This paper presents a new design for the core of a span-morphing unmanned aerial vehicle (UAV) wing that increases the spanwise length of the wing by fifty percent. The purpose of morphing the wingspan is to increase lift and fuel efficiency during extension, to increase maneuverability during contraction, and to add roll control capability through asymmetrical span morphing. The span morphing is continuous throughout the wing, which is comprised of multiple partitions. Three main components make up the structure of each partition: a zero Poisson's ratio honeycomb substructure, telescoping carbon fiber spars and a linear actuator. The zero Poisson's ratio honeycomb substructure is an assembly of rigid internal ribs and flexible chevrons. This innovative multi-part honeycomb design allows the ribs and chevrons to be 3D printed separately from different materials in order to offer different directional stiffness, and to accommodate design iterations and future maintenance. Because of its transverse rigidity and spanwise compliance, the design maintains the airfoil shape and the cross-sectional area during morphing. The telescoping carbon fiber spars interconnect to provide structural support throughout the wing while undergoing morphing. The wing model has been computationally analyzed, manufactured, assembled and experimentally tested.

Keywords: airfoil; 3D printing; carbon fiber tubes; telescoping spars; chevrons

---

1. Introduction

The aerodynamic properties of conventional airplane wings can be altered using flaps, slats, and ailerons on the exterior of the wing. These components serve to alter the geometry and surface area of the wing, but their use is accompanied by an unwanted drag and thus a lower fuel efficiency. Conventional wings accomplish only limited mission objectives and cannot be optimized for the entire flight envelope [1]. Morphing wings, in contrast, can change their geometric configurations at different points of the flight envelope to significantly improve the aerodynamic performance at various flight conditions in a variety of missions and functions [2]. However, the need to develop a wing structure that is stiff enough to carry aerodynamic loads and, at the same time, flexible enough to morph, poses many design challenges, especially if a seamless skin is sought.

The optimal wingspan required at different flight conditions is different from the fixed wingspan selected in a traditional aircraft design based on the cruise condition. Wings with large span have good range and fuel efficiency but have relatively low cruise speeds and lack maneuverability. Aircrafts with low aspect ratio wings, on the other hand, can fly faster and can become more maneuverable, but show poor aerodynamic efficiency [3]. Hence, wings with an adjustable wingspan are capable of improving the overall flight performance. Changing the wingspan directly alters the surface area of the wing, which affects the lift generated by the wing. Adjusting the span length, while maintaining a constant airfoil cross-sectional area during flight, enables the optimization of endurance, maneuverability, stability, fuel efficiency, and roll control at any specific flight condition [4]. Span morphing can be done symmetrically, where both wings are morphed identically and simultaneously, or asymmetrically, where the morphing of one wing is different from the other one. Asymmetric morphing would allow for roll control without the need for traditional ailerons [4]. The variety of modes in span length morphing introduces added benefits, due to their adaptability to flight conditions and mission profiles.

Dramatic wing configuration changes have been explored [2]. The oldest and most common concept of wing span morphing is the telescoping wing design. An example of such a design was developed by Blondeau and Pines [5]. Telescoping wings do not need flexible skin, but are usually heavier and more complex than other designs. In addition, telescoping usually takes place at the outer surface of the wing leading to discontinuities or a wing with a nonuniform cross section. Designing a successful and practical span-morphing wing requires developing a lightweight effective wing core with an actuation system, and a compliant wing skin that maintains its shape at all morphing configurations. The core should support the skin, carry the aerodynamic loads, and maintain the airfoil shape during morphing. The skin should have a proper directional stiffness that can deform with morphing without producing significant Poisson's ratio effect and should maintain proper tautness at all morphing configurations.

Different core designs for span morphing wings have been presented. Vocke III et al. [6] presented a core with a zero Poisson's ratio that can extend and slide over carbon fiber spars. Ajaj et al. [4] introduced the idea of the Zigzag wingbox that is composed of various morphing partitions, where each partition has some hinged beams that change the span of the partition upon actuation. Woods and Fristwell [7] introduced the Adaptive Aspect Ratio (AdAR) concept that includes a telescopic rectangular box spar, sliding ribs and a strap driven system for actuation. Ajaj et al. [8] developed a span morphing wing with a compliant spar. This wing is also divided into multiple partitions, where the spar in each partition is made of a series of compliant joints in the form of concentric overlapping tubes that can change the span of the partition. Ajaj et al. [9] presented the idea of the Gear driveN Autonomous Twin Spar (GNATSpar) wing for a mini-unmanned aerial vehicle (UAV). The spar of each side of the wing in this design extends to the opposite side through a rack-and-pinion mechanism to vary the wingspan length. Each of the aforementioned ideas comes with its own design challenges and complexities. Accordingly, Ajaj and Jankee, in their recent work [10], designed, built and flight-tested a span-morphing UAV, but relied on the old, well-established telescoping concept.

Span morphing wings require not just flexible skin, but stretchable skin that can significantly expand without buckling or producing an excessive Poisson's-ratio effect. Multiple studies presented designs and fabrication details of flexible/stretchable skins with a near-zero Poisson's ratio, for morphing wing applications [11–14]. La et al. [15] recently presented a survey on the status and challenges of designing skins for aircrafts with morphing wings. Ajaj et al. [9] used flexible elastomeric skin on their GNATSpar wing design with 5% pre-tension. This 0.5 mm-thick latex skin increased the required actuation force and had a significant Poisson's effect that resulted in a nonuniform airfoil shape along the span. Hence, it was suggested that flexible skins with a lower Young's modulus, such as Tecoflex and Rhodorsil V-330/CA-35 Silicone elastomers, be used to improve the skin behavior. Bubert et al. [12] and Vocke III et al. [6] developed an elastomeric-matrix-composite (EMC) skin made of Rhodorsil V-330/CA-35 silicone rubber with embedded chordwise unidirectional carbon fibers. This skin had a near-zero Poisson's effect. Recently, Jakubinek et al. [16,17] developed a stretchable skin based on carbon nanotube-polyurethane sheets with approximately 25 wt% carbon nanotubes.

This paper presents an improved design for the core structure of the span-morphing wing presented in Reference [6]. The new design features an assembled honeycomb substructure made of flexible chevrons, integrated with rigid internal ribs. This results in spanwise compliance to enable morphing, with a transverse rigidity to effectively carry the aerodynamic loads. This is in contrast to the design in Reference [6] that was made of one flexible material, and hence lacked the benefit of having varied directional stiffness. The new honeycomb substructure is also much easier to manufacture than that reported in Reference [6], since it eliminates the need for 3D-printed support structures, if a fused deposition modeling (FDM) 3D printer is used. It also avoids all complications that come from printing one large complicated structure in a single 3D printing job. In addition, telescoping carbon fiber tubes are used to replace the idea of sliding ribs on a fixed spar, presented in Reference [6]. The telescoping action is done in each partition, hence effectively and efficiently utilizing the precious space inside the wing. The model in Reference [6] did not include any actuators. The current work uses linear actuators in the core of each partition to morph the wing. Structural analyses of the span-morphing partition have been performed to ensure the ability of the wing to carry the aerodynamic loads without significant transverse deformations, and to assess the ability of the linear actuators to generate the required forces to actuate the wing partitions in both directions. The proposed design has been manufactured, assembled and tested. The testing results qualitatively confirmed all computational simulations and proved the effectiveness of the new design. Appendix A presents some of the aforementioned designs in the literature, some drawbacks based on the authors' opinion, and the improvements that the proposed design offers in order to avoid such drawbacks.

The rest of this paper is organized as follows: Section 2 introduces the new design of the wing core; Section 3 presents the structural analyses performed and the computational results; Section 4 addresses some manufacturing details; Section 5 covers testing. This is followed by a final summary and conclusion.

2. Wing Core Design

The design requirements for the morphing wing core were: (1) change the wingspan by 50% from the compressed state to the expanded state, (2) maximize the transverse and chordwise stiffness while keeping the spanwise stiffness low to enable morphing, (3) support a seamless, stretchable skin without changing the airfoil shape along the wing span, (4) use lightweight materials for the wing load-carrying structure and ensure no failure or large deformations to this structure under the expected aerodynamic loads. The UAV wing shape is rectangular, with a chord length of 27 cm and baseline (non-morphed) span length of 102 cm. Compressed and extended span lengths of 81.6 cm and 122.4 cm, respectively, were chosen to reflect a fifty percent span length increase from the compressed configuration. The NACA 0015 standard symmetrical airfoil was chosen for its high lift-to-drag ratio. The wing comprises five partitions, as can be seen in Figure 1. Each partition can be compressed or extended as shown in Figure 2.

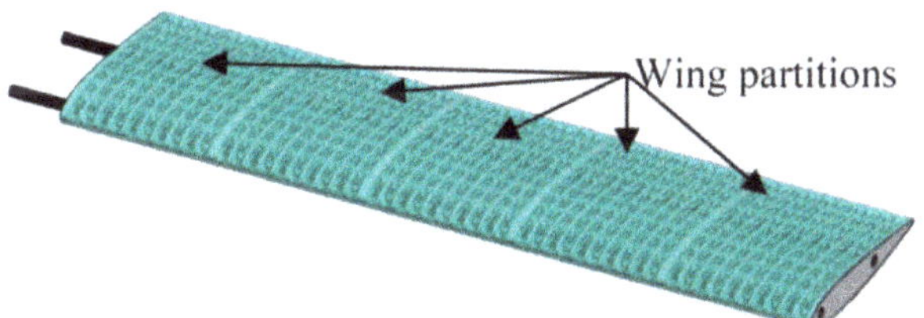

Figure 1. The proposed span morphing wing assembly.

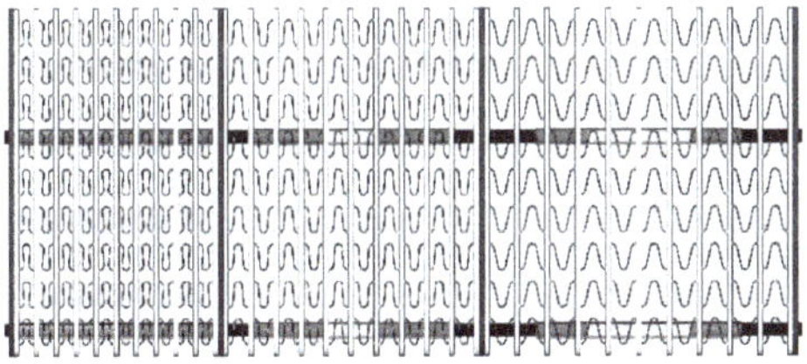

Figure 2. Three partitions: compressed, baseline and extended.

Three main components make up the structure of each partition, as shown in the cutaway view in Figure 3: A zero Poisson's ratio honeycomb substructure, two telescoping carbon fiber spars, and a linear actuator. The zero Poisson's ratio honeycomb substructure allows for a fifty percent extension while maintaining the airfoil shape and the cross-sectional area. The telescoping carbon fiber tubes interconnect to provide the main structure of the wing. Linear actuators in all partitions provide the means of actuating the entire wing.

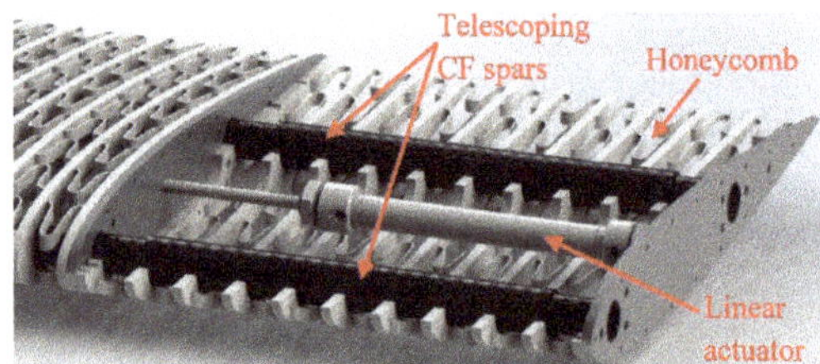

Figure 3. Cutaway view of a single partition.

The whole wing can be covered by a seamless, continuous, pre-tensioned and stretchable latex skin or composite skin of silicone rubber matrix with embedded unidirectional chordwise carbon fibers. This skin allows for spanwise extension while maintaining chordwise stiffness. Along with the honeycomb substructure, the chordwise orientation of the carbon fibers in the skin ensures a near-zero Poisson's ratio while morphing occurs. Similar skin designs have been developed by Bubert et al. [12] and Chen et al. [14]. Compared to the designs in References [4,9] (figures in Appendix A), the proposed design avoids the significant nonuniformity of the airfoil shape along the span because of the presence of internal ribs and chevrons to support the skin between the main ribs.

2.1. Assembled Honeycomb Substructure

The honeycomb substructure, presented for the first time in Reference [6], was 3D printed of acrylic-based photopolymer as one piece. This was replaced in the proposed design by an assembly of flexible interconnecting beams or "chevrons" made of Nylon Alloy 910, connected to rigid internal ribs made of polylactic acid (PLA) as shown in Figure 4 (left). Because of the rigid internal ribs, this modified design has a much stronger transverse and chordwise rigidity to support the skin and carry the aerodynamic loads without significant transverse deformations, while still featuring the spanwise compliance provided through the flexible chevrons. The design maintains the constant airfoil shape and cross-sectional area, which is not guaranteed in the flexible design in Reference [6]. Figure 4 (right) shows the internal rib design. The chevron design was also modified to take the shape of a cosine wave, which reduces the stress concentrations that can develop around the sharp curves of the honeycomb design in Reference [6]. The honeycomb substructure is fastened to the main spar/rib structure of the wing as shown in Figure 5.

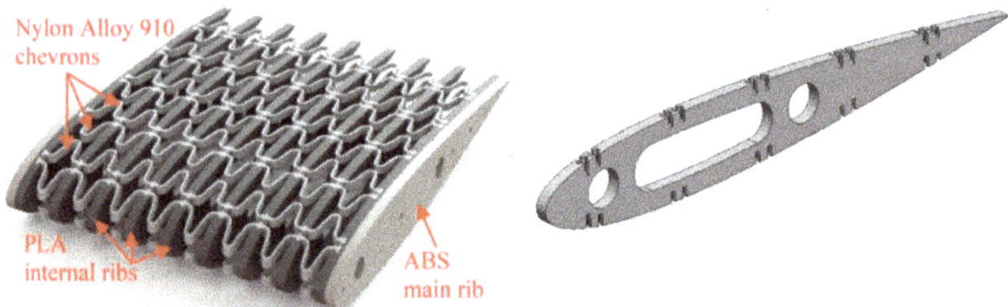

Figure 4. (**left**) The honeycomb substructure, (**right**) internal rib design.

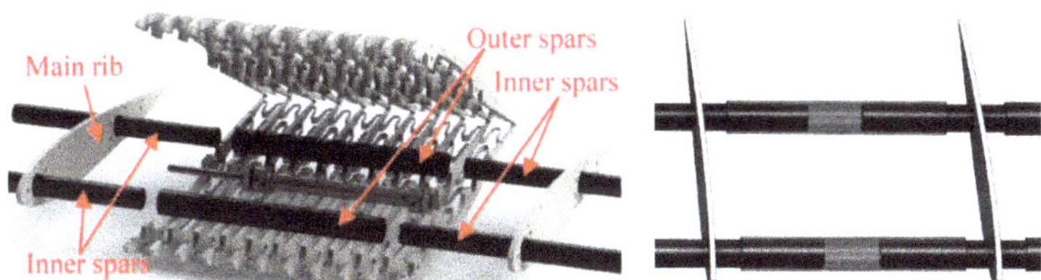

Figure 5. (**left**) Exploded view of a wing partition, (**right**) interconnected telescoping spars.

2.2. Interconnected Telescoping Spars

The spar assembly provides structural rigidity and resists bending of the wing under aerodynamic loads. In contrast to past designs which relied on either a traditional single carbon fiber spar, with telescoping taking place within the fuselage, or spars that fold within the morphing partition [4,6,18], the proposed design relies on interconnected, telescoping carbon fiber spars, with the telescoping action taking place within each partition of the wing. The benefit of this design is that it keeps the actuation within the individual morphing partitions, thus utilizing the space inside the wing more efficiently and providing more support to the honeycomb substructure than what the zigzag concept offers [4], for example.

The central outer spar is attached to the central internal rib of the honeycomb to prevent unwanted side-to-side motion. The inner spars pass through and are bonded to the main ribs, as shown in the exploded view in Figure 5. In order to keep the spars from moving too far apart and the center spar from becoming dislodged, the linear actuators limit the ultimate extension and contraction of the system. This allows the system to achieve the motion necessary for morphing, while locking the structure in place for both the compressed and extended configurations. A layer of Teflon between the inner and outer spars is used to minimize the sliding friction.

The carbon fiber tubes are made of layers of interwoven T300 carbon fibers. The main ribs, 3D printed of acrylonitrile butadiene styrene (ABS), are bonded to the inner carbon fiber hollow spars. These inner spars have a total length of 16.7 cm, an outer diameter of 13 mm and inner diameter of 11 mm. Teflon tubes with an outer diameter of 15 mm are affixed to the outside of the inner spars. The outer spars have an outer diameter of 18 mm, inner diameter of 16 mm and length of 16.3 cm.

3. Simulations

To analyze the effect of the aerodynamic loads on the major components of the proposed design, SOLIDWORKS Simulation (version 2017 by Dassault Systemes, Vélizy-Villacoublay, France) was used to perform a finite element analysis (FEA) on the developed model. A static analysis was performed on the spars to determine the maximum displacement under aerodynamic loads. First, an elliptical load distribution was assumed over the wing as shown in Figure 6.

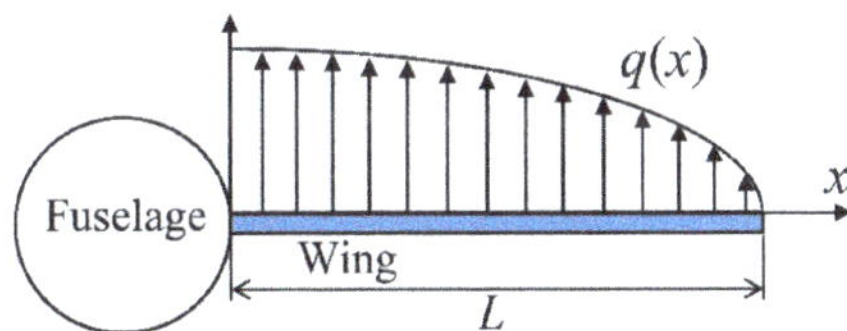

Figure 6. Distribution of aerodynamic loads over a single wing.

Equation (1) represents the distribution of the load, $q(x)$, the wing experiences based on the weight of the aircraft and the weight of the wing [19].

$$q(x) = \frac{2nW_{total}\sqrt{L^2 - x^2}}{\pi L^2} - nW_{wing}, \tag{1}$$

W_{total} is the total weight of the UAV (assumed to be 15 Kg, which is an average weight of a typical UAV of the same size), W_{wing} is the weight of the wing (assumed to be $W_{total}/8$, which is a commonly used estimate for the weight of the wing), n is the load factor (defined as lift/total weight, and taken to be 1.5 here), L is the span length, and x is the position measured from the wing root along the span. The load can be discretized and broken down into six individual distributed loads to be applied to the six main ribs for analysis. Since the loads were transferred from the skin and the honeycomb substructure to the spars via the six main ribs, only the six ribs and the interconnected spars were used in this simplified bending analysis. Each rib had two spars running through it, as shown in Figure 7. The reaction forces between the ribs and the spars were analyzed, showing that the rear and front spars take 2/3 and 1/3 of the total load, respectively. Hence, another analysis was conducted on the rear spar separately, where the root of the spar was constrained, and two-thirds of the total load was applied at the locations of the six main ribs. Split lines were created in SOLIDWORKS to simulate where the ribs are located on the spars. A convergence analysis was performed on both the compressed and extended configurations. The maximum displacement occurred at the tip of the wing for both configurations, as anticipated. The maximum displacement of the spar assembly was 7 mm and 15 mm for the compressed and extended configurations, respectively, with large strength ratios.

Figure 7. Static analysis on the rib-spar assembly (loads are applied at rib locations).

Multiple studies were also conducted on the honeycomb substructure. First, a static study was performed in order to find the force required to expand the honeycomb substructure through the full stroke. This was done to ensure that the linear actuators would be able to supply the force required to actuate the honeycomb substructure. One end of the honeycomb was fixed while the opposite end was given a 4 cm-prescribed displacement to expand or compress the structure. The reaction force in the direction of the actuation was found to be 8.05 N, which the linear actuators are more than capable of delivering (L12-R Actuonix Linear Electric Actuator [20] provides 42 N output force). Figure 8 displays the constraints of the honeycomb substructure in the FEA simulation, along with the prescribed displacements and reaction forces.

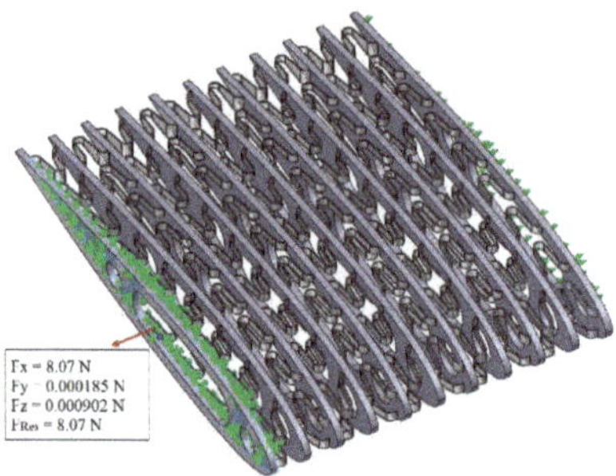

Figure 8. Fixed constraints, a 4 cm prescribed displacement in the longitudinal direction and the resultant reaction forces on the honeycomb substructure.

The honeycomb substructure and the internal ribs act as the main support for the skin of the wing. Hence, another study was conducted to check how the structure reacts to a large vertical load while in the extended configuration. The honeycomb was simulated by removing the spar supports to eliminate the rigid structure that would resist the vertical deflection. The magnitude of the vertical load was 100 N and was applied to the top half of the structure while a 4 cm-prescribed displacement was applied to the end of the honeycomb section. This large force was chosen to simulate an extreme situation, with a greater vertical load than the wing would experience in real life. One of the end ribs and the middle internal rib were fixed, simulating the honeycomb in the assembly.

The converged mesh that was used consisted of 1,077,564 nodes equivalent to 615,139 second-order tetrahedral elements, created using the curvature-based mesher in SOLIDWORKS Simulation. Despite the large vertical load, the maximum stress was well below the yield strength of Nylon Alloy 910, with a 5.3 minimum factor of safety. The maximum displacement was 4.5 mm. This means that the deformations of the chevrons will not interfere with the linear actuators or the internal spars. Figure 9 shows the deformed and stressed structure.

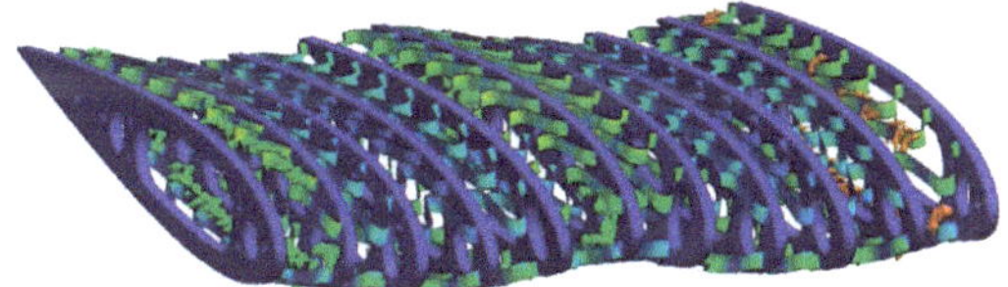

Figure 9. The static study on the honeycomb substructure with a 100 N load and a 4 cm-prescribed displacement applied (maximum displacement = 4.5 mm, and minimum factor of safety = 5.3).

4. Manufacturing

Because of the complex geometry of the honeycomb, 3D printing was determined to be the only viable option for manufacturing. The internal ribs and chevrons were 3D printed separately from PLA and Nylon Alloy 910, respectively. Each row of chevrons was 3D printed as one piece, yielding 20 rows of separately 3D printed chevrons per honeycomb partition. The honeycomb substructure was assembled afterwards with a tight press fit between the ribs and chevrons as shown in Figure 10.

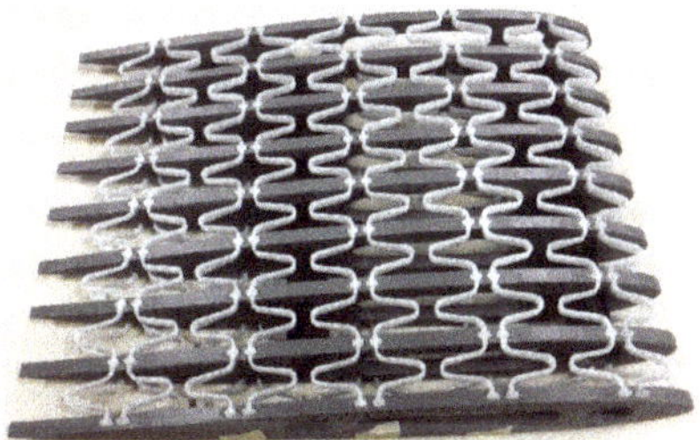

Figure 10. The assembled honeycomb substructure.

This new design allowed the parts to be 3D printed at a much faster rate compared to that in [6], because of the elimination of the 3D printing support material. The selected materials achieved the correct combination of flexibility in the chevrons to allow for spanwise morphing and rigidity in the ribs to eliminate transverse and chordwise deformations. Hence, it ensures a uniform cross section throughout the wing span.

5. Testing

To test the spanwise compliance of the assembled partition, the structure was oriented with the ribs parallel to the floor, known weights were hung from the partition to induce extension, and the resulting displacements were measured. Similar tests were conducted with the partition under compression as shown in Figure 11.

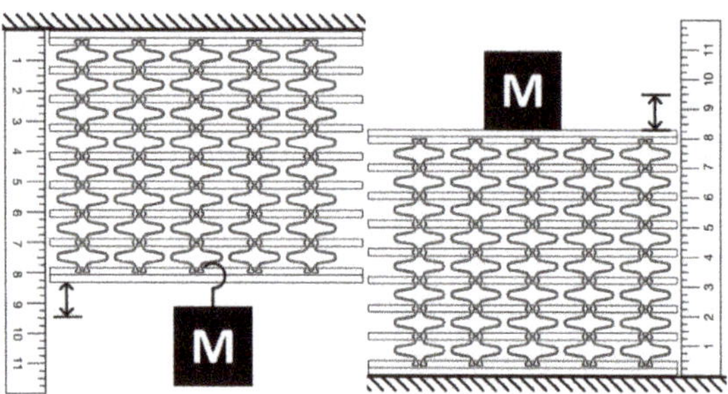

Figure 11. Testing the compliance of the assembled structure.

The span morphing partition is shown assembled and under compression in Figure 12. Figure 13 shows the results of the honeycomb compliance tests. The relation between displacement, *d*, and input force, *F*, was nearly linear in both cases, and can be expressed as

$$d = \begin{cases} 0.1498F - 0.6156\,(\text{cm})\,\text{Extension} \\ 0.1172F + 0.3722\,(\text{cm})\,\text{Compression} \end{cases}, \tag{2}$$

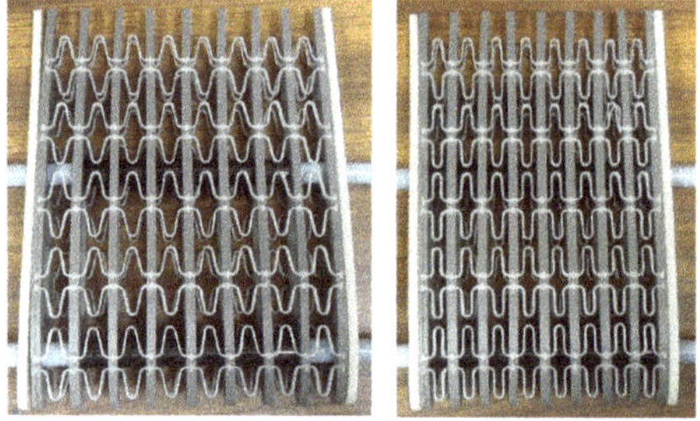

Figure 12. The span morphing partition (**left**) assembled, (**right**) under compression.

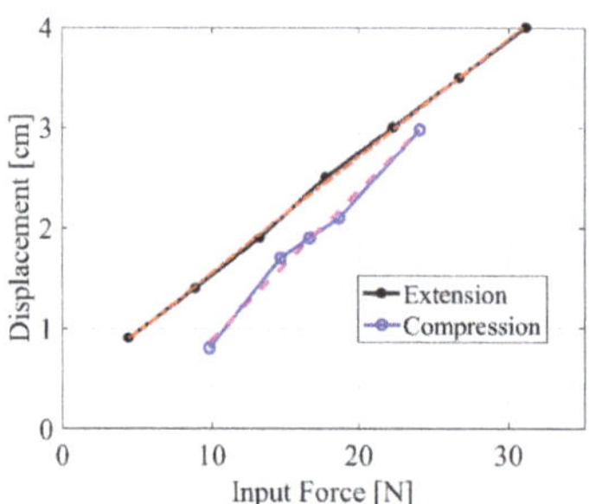

Figure 13. Honeycomb displacement vs. input force, in extension and compression tests.

The stiffness of the honeycomb during extension was approximately 6.67 N/cm while the stiffness during compression was approximately 8.53 N/cm. Based on these experimental results, the geometry and the material of the chevrons can be iterated to accommodate any changing design requirements.

The linear actuators were able to fully extend and compress the manufactured wing partition smoothly, as shown in Figure 14. Fixing one end rib of the wing partition in a horizontal position and hanging several weights on the opposite end rib, to induce bending moments equivalent to the expected aerodynamic bending loads, did not prevent the actuator from fully extending and compressing the model.

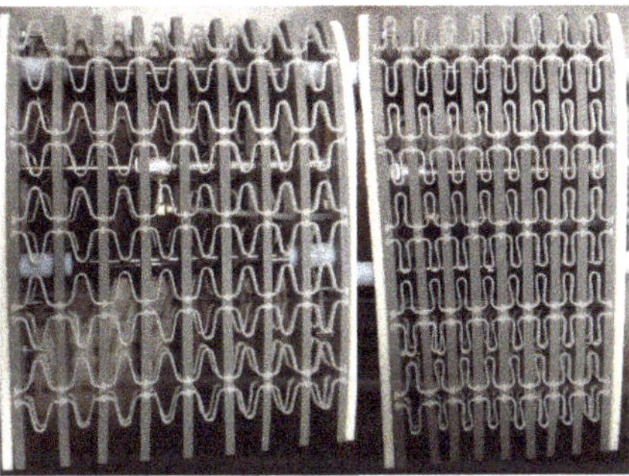

Figure 14. The actuation of the assembled span morphing wing partition (**left**) extended, (**right**) compressed.

Current work is focusing on covering the proposed core design by a seamless pre-tensioned latex skin, as shown in Figure 15, and performing wind tunnel tests. The skin is bonded to the internal ribs and will remain under tension in both actuation directions. This skin is the most suitable for the proposed design because when the wing section expands, the in-plane stiffness of the skin increases, allowing it to effectively transfer the applied aerodynamic loads to the ribs and spars. When the wing section contracts, the skin will remain under tension because of the applied pre-tension and the support of the chevrons and the internal ribs. The chevrons that are compressed between the internal ribs in this case will compensate for the decrease in skin stiffness by providing support to the skin.

Figure 15. Pre-tensioned latex skin covering a wing partition.

6. Conclusions

This paper presented a new and improved design of a span-morphing UAV wing core model, where an assembled honeycomb substructure with interconnected telescoping spars and a linear actuator form one partition in a five-partition wing, able to make a fifty-percent span extension from a compressed configuration. The honeycomb substructure, made of rigid internal ribs with flexible chevrons, provides the necessary spanwise compliance to allow for actuation, and transverse rigidity to carry the aerodynamic loads, while maintaining the airfoil shape. The interconnected spars are the main load-carrying members in the design, and effectively enable the telescoping action to happen inside each partition of the wing. The linear actuators have the necessary output force to drive the system and can keep the partition at any desired configuration for any period of time. The new core is believed to have multiple advantages over previously published designs and improves the viability of a honeycomb-based span-morphing concept. The telescoping spar concept presented here resolves the drawbacks related to discontinuity and nonuniformity in the traditional telescoping wing designs, by bringing the telescoping mechanism inside the wing, rather than having it at the outer surface of the wing. This results in a seamless, continuous and uniform span-morphing wing. By separately 3D printing the internal ribs and chevrons of the honeycomb, different materials can be selected for the ribs and chevrons based on the design requirements, and components can be replaced easily during maintenance. Future work will focus on performing optimization studies on the wing with a stretchable skin, wind tunnel testing, failure analysis and reliability analysis.

Author Contributions: Conceptualization, P.L.B. and E.B.; methodology, E.B., A.A., R.P. and K.S.; software, E.B. and K.S.; validation, P.L.B. and R.P.; writing—original draft preparation, E.B., A.A. and K.S.; writing—review and editing, P.L.B.; visualization, E.B. and P.L.B.; supervision, P.L.B.; project administration, P.L.B.; funding acquisition, P.L.B.

Funding: This research was funded by the Instructionally Related Activities (IRA) grant at California State University, Northridge, CA, USA.

Acknowledgments: This work is a part of the research-based "Smart Morphing Wing" senior design project founded by the first author. The support of the Mechanical Engineering Department, the Instructionally-Related Activities (IRA) grant, and the Student Travel and Academic Research (STAR) grant at California State University, Northridge (CSUN) are acknowledged.

Conflicts of Interest: The authors declare no conflicts of interest.

Appendix A

Table A1. Drawbacks of previous span-morphing wing designs, based on the authors' opinion, and the improvements made in the proposed design.

Model	Drawbacks	Improvements in the Proposed Model
Vocke III et al. [6]	-The whole section is made of a single, flexible material, hence there is no directional stiffness. -The whole section is 3D printed in a single job (that will need a support structure if FDM 3D printer is used). -No actuators reported.	-The section is divided into internal rigid ribs that are connected to flexible chevrons. The spanwise stiffness is significantly lower than the transverse and chordwise stiffness because of the different materials used. -Internal ribs and chevrons are 3D printed separately and easily assembled. -Linear actuators are used.
Ajaj et al. [4]	-Six joints in each partition. -No support to the skin. -Morphing (zigzag) mechanism occupies a lot of space.	-No joints at all in each partition. -The internal ribs and the chevrons support the skin and maintain the airfoil shape along the span. -Minimum space is required since the telescoping action is happening inside the spars.
Ajaj et al. [9]	-No support to the skin between the main ribs, resulting in a significant Poisson's effect and nonuniform airfoil shape.	-The internal ribs and the chevrons support the skin and maintain the airfoil shape along the span.
Samuel and Pines [5]		
Ajaj et al. [10]	-Nonuniform airfoil shape along the span (telescoping is happening at the outer surface of the wing).	-Uniform airfoil shape along the span (telescoping is happening internally inside the spars).

References

1. Min, Z.; Kien, K.V.; Richard, L.J.Y. Aircraft morphing wing concepts with radical geometry change. *IES J. Part A Civ. Struct. Eng.* **2010**, *3*, 188–195. [CrossRef]
2. Sofla, A.Y.N.; Meguid, S.A.; Tan, K.T.; Yeo, W.K. Shape morphing of aircraft wing: Status and challenges. *Mater. Des.* **2010**, *31*, 1284–1292. [CrossRef]
3. Anderson, J.D. *Fundamentals of Aerodynamics*, 5th ed.; McGraw-Hill: New York, NY, USA, 2011.
4. Ajaj, R.M.; Flores, E.I.S.; Friswell, M.I.; Allegri, G.; Woods, B.K.S.; Isikveren, A.T.; Dettmer, W.G. The Zigzag wingbox for a span morphing wing. *Aerosp. Sci. Technol.* **2013**, *28*, 364–375. [CrossRef]
5. Samuel, J.B.; Pines, D. Design and testing of a pneumatic telescopic wing for unmanned aerial vehicles. *J. Aircr.* **2007**, *44*, 1088–1099. [CrossRef]
6. Vocke, R.D., III; Kothera, C.S.; Woods, B.K.S.; Wereley, N.M. Development and testing of a span-extending morphing wing. *J. Intell. Mater. Syst. Struct.* **2011**, *22*, 879–890. [CrossRef]
7. Woods, B.K.S.; Friswell, M.I. The adaptive aspect ratio morphing wing: Design concept and low fidelity skin optimization. *Aerosp. Sci. Technol.* **2015**, *42*, 209–217. [CrossRef]
8. Ajaj, R.M.; Flores, E.I.S.; Friswell, M.I.; Diaz de la O., F.A. Span Morphing Using the Compliant Spar. *J. Aerosp. Eng.* **2015**, *28*, 04014108. [CrossRef]
9. Ajaj, R.M.; Friswell, M.I.; Bourchak, M.; Harasani, W. Span morphing using the GNATSpar wing. *Aerosp. Sci. Technol.* **2016**, *53*, 38–46. [CrossRef]
10. Ajaj, R.M.; Jankee, G.K. The transformer aircraft: A multimission unmanned aerial vehicle capable of symmetric and asymmetric span morphing. *Aerosp. Sci. Technol.* **2018**, *76*, 512–522. [CrossRef]
11. Olympio, K.R.; Gandhi, F. Zero Poisson's ratio cellular honeycombs for flex skins undergoing one-dimensional morphing. *J. Intell. Mater. Syst. Struct.* **2010**, *21*, 1737–1753. [CrossRef]
12. Bubert, E.A.; Woods, B.K.S.; Lee, K.; Kothera, C.S.; Wereley, N.M. Design and fabrication of a passive 1D morphing aircraft skin. *J. Intell. Mater. Syst. Struct.* **2010**, *21*, 1699–1717. [CrossRef]
13. Gong, X.; Huang, J.; Scarpa, F.; Liu, Y.; Leng, J. Zero Poisson's ratio cellular structure for two-dimensional morphing applications. *Compos. Struct.* **2015**, *134*, 384–392. [CrossRef]
14. Chen, J.; Shen, X.; Li, J. Zero Poisson's ratio flexible skin for potential two-dimensional wing morphing. *Aerosp. Sci. Technol.* **2015**, *45*, 228–241. [CrossRef]
15. La, S.; Alsaidi, B.; Joe, W.Y.; Akbar, M. Survey of skin design for morphing wing aircraft: Status and challenges. In Proceedings of the AIAA Aerospace Sciences Meeting, Kissimmee, FL, USA, 8–12 January 2018.
16. Jakubinek, M.; Ashrafi, B.; Martinez-Rubi, Y.; Laqua, K.; Palardy-Sim, M.; Roy, S.; Sunesara, A.; Rahmat, M.; Denommee, S.; Simard, B. Multifunctional skin materials based on tailorable, carbon-nanotube-polyurethane composite sheets. In Proceedings of the 2018 AIAA/ASME/ASCE/AHS/ASC Structures, Structural Dynamics, and Materials Conference, Kissimmee, FL, USA, 8–12 January 2018. [CrossRef]
17. Jakubinek, M.; Roy, S.; Palardy-Sim, M.; Ashrafi, B.; Shadmehri, F.; Renaud, G.; Barnes, M.; Martinez-Rubi, Y.; Rahmat, M.; Simard, B.; et al. Stretchable structure for a benchtop-scale morphed leading edge demonstration. In Proceedings of the AIAA Scitech 2019 Forum, San Diego, CA, USA, 7–11 January 2019. [CrossRef]
18. Ajaj, R.M.; Friswell, M.I.; Flores, E.I.S.; Little, O.; Isikveren, A.T. Span morphing: A conceptual design study. In Proceedings of the 53rd AIAA/ASME/ASCE/AHS/ASC Structures, Structural Dynamics, and Materials Conference, Honolulu, HI, USA, 23–26 April 2012.
19. Doherty, D. Analytical modeling of aircraft wing loads Using MATLAB and Symbolic Math Toolbox. *MATLAB Dig.* 2009. Available online: https://www.mathworks.com/company/newsletters/articles/analytical-modeling-of-aircraft-wing-loads-using-matlab-and-symbolic-math-toolbox.html (accessed on 3 January 2019).
20. Actuonix Rod Actuators. Available online: https://www.actuonix.com/Rod-Actuators-s/1924.htm (accessed on 3 January 2019).

Article

Exploring an AM-Enabled Combination-of-Functions Approach for Modular Product Design

Charul Chadha [1], Kathryn A. Crowe [2], Christina L. Carmen [3] and Albert E. Patterson [4],*

[1] Department of Aerospace Engineering, University of Illinois at Urbana-Champaign, 306 Talbot Laboratory, 104 South Wright Street, Urbana, IL 61801, USA; charulc2@illinois.edu

[2] NASA Marshall Space Flight Center, Huntsville, AL 35811, USA; kathryn.crowe@nasa.gov

[3] Department of Mechanical and Aerospace Engineering, University of Alabama in Huntsville, Technology Hall N274, 300 Sparkman Drive, Huntsville, AL 35899, USA; christina.carmen@uah.edu

[4] Department of Industrial and Enterprise Systems Engineering, University of Illinois at Urbana-Champaign, 117 Transportation Building, 104 South Mathews Avenue, Urbana, IL 61801, USA

* Correspondence: pttrsnv2@illinois.edu; Tel.: +1-217-333-2731

Received: 21 September 2018; Accepted: 14 October 2018; Published: 16 October 2018

Abstract: This work explores an additive-manufacturing-enabled combination-of-function approach for design of modular products. AM technologies allow the design and manufacturing of nearly free-form geometry, which can be used to create more complex, multi-function or multi-feature parts. The approach presented here replaces sub-assemblies within a modular product or system with more complex consolidated parts that are designed and manufactured using AM technologies. This approach can increase the reliability of systems and products by reducing the number of interfaces, as well as allowing the optimization of the more complex parts during the design. The smaller part count and the ability of users to replace or upgrade the system or product parts on-demand should reduce user risk, life-cycle costs, and prevent obsolescence for the user of many systems. This study presents a detailed review on the current state-of-the-art in modular product design in order to demonstrate the place, need and usefulness of this AM-enabled method for systems and products that could benefit from it. A detailed case study is developed and presented to illustrate the concepts.

Keywords: additive manufacturing; modular design; design-for-manufacturability; design optimization; part consolidation; product re-design

---

1. Introduction

As additive manufacturing (AM) technology becomes more widely used and accepted within the engineering and production world, the many benefits it offers are becoming increasingly useful in engineering design. One of the main advantages offered is the elimination of tooling and work fixtures needed in traditional manufacturing processes [1–4]. AM builds parts in layers directly from computer-aided design (CAD) data with few geometric restrictions, allowing the use and manufacturing of parts with very complex features. This design freedom is very useful in the production of optimized parts which could replace existing single parts or even several whole parts which interact with each other. Formally, the design goal is to either consolidate several parts into one or to decompose and recombine parts into new ones [4–8]. Ideally, these new parts are functionally superior to the original ones; however, just having more control over the part geometry during processing may provide many benefits on its own by simplifying the manufacturing process [9–11]. The variety of new and tailored materials available for AM processing could also allow the use of materials which would be infeasible during more traditional design and manufacturing processes. A good example of this is the recent widespread use of titanium alloys in the manufacturing of small and complex medical devices [12–16] ; most of these new designs would be impractical or infeasible to manufacture using traditional methods due to the material and geometric design problems involved in producing the devices.

Since no special tooling or fixtures are needed, it is also possible to produce replacement or upgraded parts on-demand as the user needs them [17,18]. Additive technologies could also allow the development of part families [19,20], allowing a single system to adapt to different jobs by the simple exchanging of a few essential components as needed; an effective method for accomplishing this is to produce a "parent" or "archetype" design and then modify as needed [21–23]. For many systems which can benefit from AM during design and production, fixed design of every component is not needed, as additional parts and part sets can be made on-demand, often even in the field [24–26]. Figure 1 shows the concept map of the general AM process flow. A plethora of additive processes are available, based on several different parent technologies and are able to effectively process a very wide variety of materials [17,27,28]. The process families are defined according to the state of the raw material before processing, the method of material deposition, and method of layer fusion. Consideration of the distinction between processes is important because various processes have different strengths and weaknesses for different applications [17,28,29].

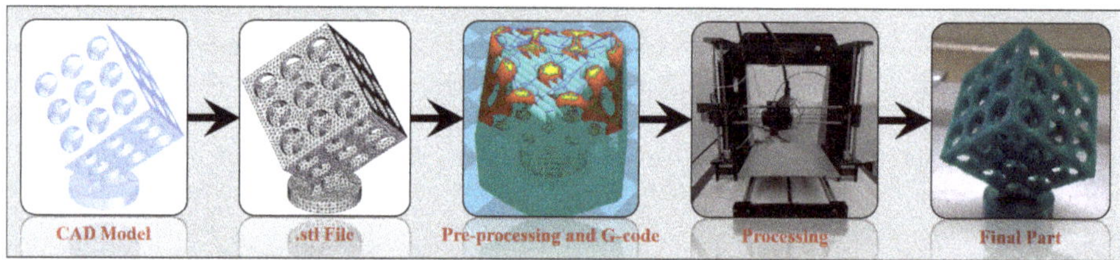

Figure 1. AM concept map.

Reduced integration complexity within many systems or products can serve to increase their reliability, giving a longer useful life and higher customer satisfaction. It is well known that most failures in any system happen at the interfaces between the system components, so the more interfaces that exist in the design, the more failure modes exist as well. The whole system reliability is also based on the reliabilities of the individual components, with the system or product reliability being based on the product of the n component reliabilities [30–32] for a system of series components (Figure 2).

$$R = P(x_1)P(x_2)P(x_3)\cdots P(x_n) \tag{1}$$

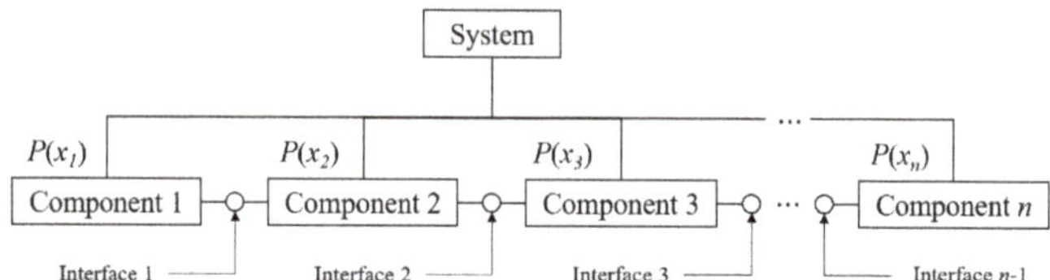

Figure 2. System reliabilities and interfaces concept.

The system can be made more or less reliable with the addition of fail-safe components and the system arrangement changes (e.g., using parallel components) [33,34]. However, the elimination of interfaces in the system should improve it by simplification in itself; reduction in the part count (i.e., fewer component reliabilities to monitor) and the careful arrangement of the components can offer further improvements to system usefulness and reliability [35–38].

The informed designer can make use of these relationships and the strengths of additive manufacturing processes during the design stage. In addition to increasing the system reliability and reducing the part count in some systems, considering design-for-AM (DFAM) principles during design can drastically expand the available design space, as well as dodge many of the pitfalls encountered during the production stage. Preventing poor manufacturability of features, preventing material problems, and addressing other manufacturing concerns are a common theme for most design-for-manufacturing (DFM) methods [11,39]. When the system under design is a "modular" system which consists of a number of subsystems, this is particularly important. Expansion of the design space is very useful in ensuring that the resulting system can be optimized and rendered more reliable and useful to the stakeholders of the system. Some interesting practical examples of "modular" systems from several recently-published studies are shown in Figure 3. Figure 3a is a modular building addition system which can be used to add floors and rooms to existing buildings, presented by Dind et al. [40]; similarly, Sharafi et al. [41] designed a system of foam-filled modular building panels (Figure 3b) that can be used to rapidly put up new structures to be used by disaster victims. Electrical, power storage, and communications grids are also common users of modular systems, as discussed in the work by Zhang et al. [42] and shown in Figure 3c. Modular mechanical systems are common as well, such as the robotic manipulator (Figure 3d) described in the work by Mishra et al. [43].

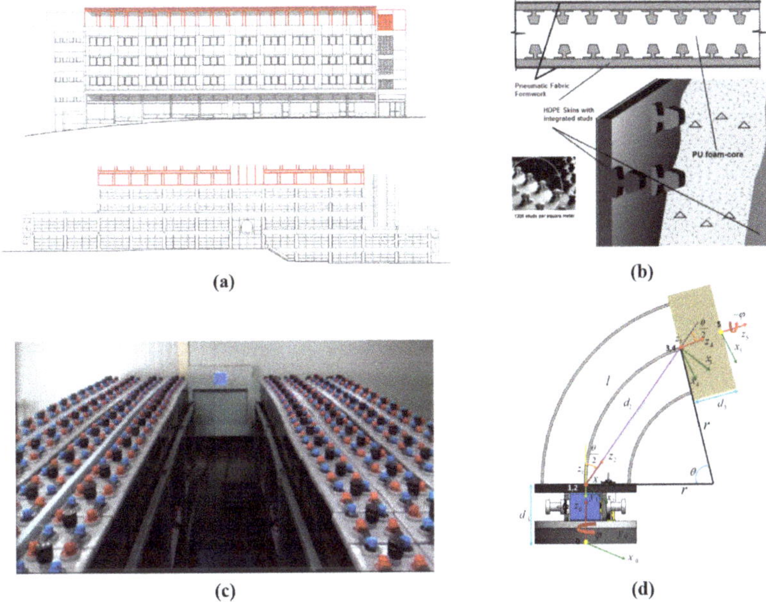

Figure 3. Modular design examples: (**a**) design method to add sections to existing buildings [40]; (**b**) modular panels for emergency housing construction [41]; (**c**) battery storage system in an electric grid [42]; and (**d**) modular mechanical manipulator [43].

Some unresolved issues exist within the DFAM paradigm, however, and restrict its applicability. First and foremost, during the design process, the practicability of actually using part consolidation or recombination in DFAM [5–7] must be carefully examined. This is not always a good approach for every system or product and this must be carefully considered when choosing to which design or re-design process to use [7,44]. Several previous studies have considered this question and have provided case studies and some guideline for using part consolidation or recombination effectively [5–7,44–47]; a more general set of guidelines for this is an important ongoing research topic. From the manufacturing perspective, AM processes tend to be at least as restrictive, if not more, in terms of manufacturability constraints than traditional processes. [29,48–50] The layer-based nature of all AM processes produce anisotropic material properties in the final parts [17,51–53] , the surface finish of final parts can be poor [17,54–56] , and the inherent residual stresses from some processes can dramatically reduce the fatigue life of the parts [57–59] . Due to these and other considerations, AM-made parts can be prone to fracture and may have poor dimensional accuracy for some processes and materials.

It should also be noted that not all materials can be effectively processed using AM, which adds a further consideration when using AM principles in design. Parts where more than one material is needed for correct functionality can be especially challenging to resolve under DFAM, as different types of materials generally need to be processed using different kinds of AM processes, which may have different levels of accuracy and material quality. Much progress has been made in all these areas, but they are important areas of concern during design when AM processes are involved. It is clear that using DFAM is a trade-off game for many systems, especially those requiring certification or which could cost human lives if they fail, and this needs to be carefully considered by designers considering the use of any DFAM method.

In the cases where AM is applicable and feasible to use in manufacturing the system and part consolidation or recombination are useful, the concepts of DFAM can be applied to modular systems; under this concept, components that were formerly designed as sub-assemblies can now be designed as single parts with many built-in functions, drastically decreasing the time and cost involved in designing, developing, integrating, and producing them. This definition of modular design by part consolidation was first described in detail by Hernandez et al. [60] in the context of rocket airframe design. The present study further explores and develops the concept for more general engineering design. The main focus of this study is to explore the idea of modular product design based on the use of large, complex, multi-function parts that can replace some sub-assemblies in modular design.

This study analyzes the concept in several sections, each presented in a section of the paper. First, an extensive review of the literature will be done to find the overall working definition of modular product design and explore traditional design modularization methods (Section 2. Next, AM-enabled design and combination-of-functions approach will be carefully defined and discussed (Section 3). The application space for the concept will also be discussed and a hybrid system proposed (Section 4). Finally, a case study to demonstrate the concept that will be developed in this study will be presented in detail and discussed (Section 5). Note that in this paper, the terms "system" and "product" are used interchangeably; in the context of this work, the only significant difference between two consists of the size and scale of the design.

2. Modular Design: Theory and Applications

The concept of modular design has been explored extensively, resulting in a wide variety of different definitions and problem-specific applications. Chung et al. [61] defined design modularity as the ability of a product to be producible via reconfiguration of existing parts or subsystems; another perspective, proposed by Gershenson et al. [62] and Baldwin et al. [63] is to intentionally design the system as a set of separable modules specific to that system or product. These definitions vary in their approach, but they both agree that the degree of modularity of a system is its ability to be separated and reconfigured. According to Bonvoisin et al. [64], modular design can be broken down into three complementary activities: (1) design with modules; (2) identification of modules; and (3) design of modules (Table 1). Concepts of "modularity" in design theory and practice have undergone several modifications in recent decades; according to an extensive literature survey conducted by Gershenson et al. [62], it is difficult to achieve a common consensus on definition and appropriate application of the terms modular or modularity. Similarly, Bonvoisin et al. [64] conclude that different researchers typically use their own interpretation of the definition and parameters used to produce modular designs, making a universal definition even more difficult to achieve. According to Todorova and Durisin [65], theories on modular management of products are often conflicted between increasing flexibility through manufacturing multiple products from few components and gaining flexibility through faster component innovation.

Table 1. Basic modular design activities.

Design with Modules	Involves designing a new product out of existing pools of pre-defined modules or parts; conceptually, this is similar to assembling a set of Lego-blocks
Identifying Modules	Requires a deep understanding of the product or system and includes studying the current product, evaluating the component grouping, clustering the components accordingly, then redesigning and integrating the defined modules and their interfaces
Design of Modules	Necessitates the design of new modules to meet new requirements or to replace obsolete or poorly-designed modules

Even with this ambiguity in definitions, the interest in modularization of product design has increased over the years because of its many benefits in some applications. Gershenson et al. [62] and Ulrich and Tung [66] studied the impact of modularity on different industries. These studies described several benefits, including:

- Decrease in production cost due to re-usability of components across different families
- Ease of product updating due to functional modules
- Increased product variety from a smaller set of components
- Decreased order lead-time due to fewer different components
- Ease of design and testing due to the decoupling of product functions
- Ease of service due to differential consumption

Gershenson et al. [62] further concluded that, because of fewer assembly types and increase utilization of part designs for several systems, production quantities for a given module can be increased, while simultaneously reducing the learning curve for production personnel. In addition, because of fewer unique parts, products can be produced in greater quantity, decreasing the production cost. The development and production cost of modular components is strongly influenced by the product architecture and optimization status. It has been suggested in recent studies that increases in modularity of designs empowers environmental friendly end-of-life design and strategy planning. Ozman [67] studied the impact on open innovation, concluding that modularity increases incentive for open innovation during final stages of the design to make it compatible with different modules. However, Ettlie and Kubarek [68] studied commercial design teams to understand the extent to which the modularity is beneficial; after closely surveying 42 companies, they concluded that novelty in design suffers after 33% of design reuse, and further innovation was absent above 55%. Kassai et al. [69] studied the impact of modularity on agile manufacturing, during which the development of a modularity matrix helped designers to better understand the product.

Based on the definition and desired impact of modularity, researchers have defined several sets of parameters that drive modularity. Ericsson and Erixon [70] defined a *modular driver* as the driving forces that impact product modularity and competitive advantages gained from using it. Through different case studies, they obtained twelve modular drivers, which cover the entire life cycle. Kreng and Lee [71] also studied this problem and identified fourteen drivers based on strategies of modular design, product market characteristics, competitive advantages, and customer requirements. Okudan et al. [72,73] categorized the methods developed for planning and using product modularization in design into three broad categories, specifically: (1) data-mining approaches; (2) mathematical approaches; and (3) design-for-X approaches.

The data mining approaches analyze information from databases and surveys concerning the functions, geometry, supply chain information, and customer feedback to come up with new designs. Agard and Kusiak [74] described a data-mining approach for modularity analysis, which defines product modules by creating function structures that define the variability and generates options to satisfy design requirements. This approach helps to bridge gap between market and designers. However, data collection may take a long time and incur high cost for the stakeholders. Some researchers also proposed reverse engineering to modify existing products using adaptive redesign as part of a data mining approach [75–77]. The mathematical modeling approach involves designing for a set of constraints and objective functions using defined variables. In this method, a design structure matrix is commonly used to represent relationships between components and objectives [78–80]. These methods are useful during product development and product planning while considering many factors. Design-for-X considers conditions for multiple views such as assembly, sustainability, and life cycle [81–83]. Typically, the X factor that is designed for is a set of defined characteristics, or even a single characteristic. It differs from typical mathematical approaches as it focuses the bulk of design resources on the optimization of a single characteristic or a small set of them. Design-for-X approaches can also be used to generate design constraints to be used for a mathematical approach to designing the system [29].

A methodology for defining modules for lifecycle engineering using an interaction matrix in weighted average sense, and using algorithms for clustering the modules was developed by Gu et al. [84]. Similarly, Ma et al. [85] developed an approach using characteristics such as cost, environmental impact, and labor time to assess the sustainability at early design stage using fuzzy algorithm. They also included end of lifecycle stage uncertainty to assess the environmental impact. Mutingi et al. [86] also proposed application of a multi criteria fuzzy grouping genetic algorithm approach during early design stage when exact evaluation criteria are unknown to model imprecise information and evaluate possible solutions. Chung et al. [61] and Kristianto and Helo [87] studied the performance of the modular structure during various stages of the supply chain, and using statistical techniques concluded that modular structure optimization can reduce the environmental impact during different stages. Yang et al. [88] and Wang et al. [89] proposed eco-design methods for life cycle engineering of electronic and electricity equipment for improving maintainability, reusability and recyclability of equipment.

Based on the review of the literature discussed in this section, a basic modular design concept relevant to practical design-for-manufacturability problems can be generated. In this paradigm, the modularization of a product or system is a design method where the system is divided into well-defined units; ideally, these units can be designed and manufactured independently from each other and then integrated together by the design teams or a dedicated team of systems engineers; a secondary goal is often to use each unit in more than one product. The overall goal is to produce more product flexibility, increase design and manufacturing efficiency, overcome manufacturing problems, improve customer usability, and decrease overall product cost [20,60,84,90]. A number of generic principles can be derived to best use and harvest value from the modular design process [60,91,92]:

1. Design of parts and subsystems should be done by specialized teams of experts
2. After design, an integration phase will be required for the system, during which the parts and sub-assemblies are converted into a usable system or product
3. Every aspect of the system should be standardized as much as possible
4. Every design technique used or derived should be developed to be as widely applicable as possible
5. Integrate further plans for testing, maintenance, and upgrades into the original design process
6. Consider end-of-life disassembly, recycling, and retirement during the design process
7. Every component and subsystem should be easily upgradable and repairable
8. Systems should be easily reconfigurable for as many missions as possible

In traditional modular product design, these principles have been best achieved by designing and building a series of independent sub-systems, each with its own mission and functionality; the sum of these subsystems constitutes the overall system or product. This concept is shown in Figure 4, where the circular connectors represent interfaces for n sub-assemblies, each with a number $(q, p, r, \ldots, s)$ of parts for a total of $q + p + r + \cdots + s$ parts. While this method is widely used in practice [64,70], it is not without its disadvantages. The main issue is the problems that naturally arise from the need to integrate the subsystems together, presumably made by different teams of designers [92]. Communication between these subsystems is key, but is not always effective due to integration issues and subsystem incompatibility. It is well known that most system reliability issues and errors occur at the interfaces within the system [32,93–95]. In order to effectively use modular design techniques in practice, a robust method of integration is needed, as discussed in the modular design literature above. The most effective systems engineering (SE) product lifecycle engines consider this in depth in their procedures. Sophisticated examples of this are the SE Engine used by the United States (US) National Aeronautics and Space Administration (NASA) [96] and System-of-Systems method used by the US Missile Defense Agency (MDA) [97].

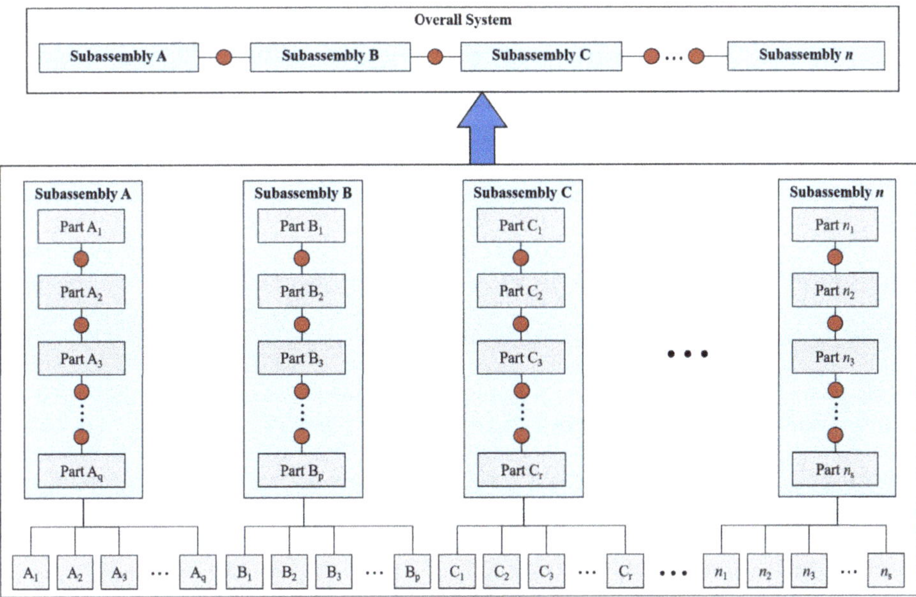

Figure 4. Modular design complexity levels (adapted from [60]).

3. AM-Enabled Modular Design (AMEMD) Concept

An effective, though not widely considered, method for modular or semi-modular design is enabled by AM technologies [4,60,98]. The ability of AM to make parts of almost unlimited complexity and with high functionality is the key [39,98]. The concept is simple: when possible, replace the subsystems within the overall system with complex parts that each have the functionality of a subsystem [60]. The benefits of modular design can still be realized, but with many fewer system interfaces and with a simpler design process. In theory, just the elimination of interfaces would dramatically improve the reliability of the system [95]. Figure 5 demonstrates the combination-of-functions concept. In the lower field, the functions are combined into a series of complex, multi-function parts which are essentially "functional modules" [60,99,100]; these are single parts but have the behavior of a sub-assembly of several parts, simply with the interfaces removed. When these are combined into the overall system, they still need to be integrated (where the red dots indicated integration points) but there are far fewer of these points than in the traditional system.

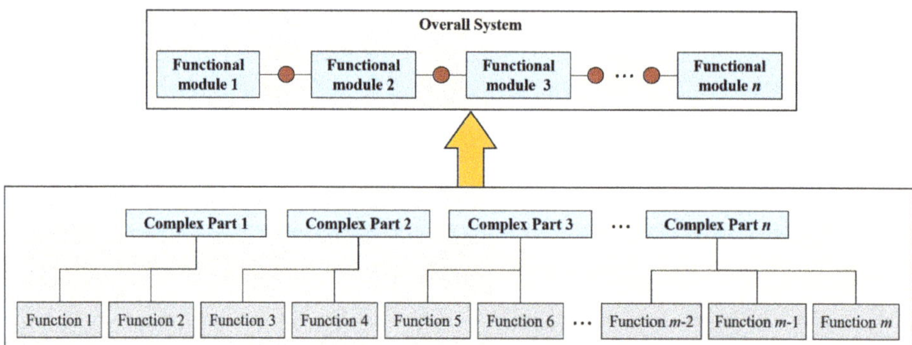

Figure 5. Combination-of-Features concept (adapted from [24,60]).

On engineering grounds, the potential objection to this approach in design is that the technique eliminates standardization and requires that the parts be highly specialized. Fortunately, one of the most important aspects of AM is its ability for easy mass customization for parts [101,102]. Since no special tooling or fixtures are required, batches of parts can be nearly infinitely customizable at the design stage [17]. If applied correctly by a knowledgeable designer, considering both the strengths and weaknesses of AM technologies, the need for a supply chain of standard off-the-shelf parts is greatly reduced or eliminated in many systems [17,24,26,28,103].

While some systems are practical to additively manufacture as a single part themselves, this effectively suppresses some of the main benefits of modular design. In particular, it is impossible to produce systems that are easily reconfigurable or reparable without replacing any major parts. It is also much more difficult to repair or maintain such systems or use them for purposes other than their core missions. When systems are built as a single part, they are usually throw-away products that are not meant to be repaired or reconfigured [60,104,105]. For most products that would benefit from modularity in design, the subsystem-level is likely the highest where it is practical to use a combination-of-functions approach. Hence the concept of feature cataloging in design [60,106], which uses a function database of options in design instead of a parts or subsystem catalog, as was often done in traditional design [107–110].

AM parts are, by definition, unique one-off or batch parts, where the cost of production depends only on the volume of material used, printing time, and post-processing requirements, as well as certification if needed. In functionally-modular systems, maintaining flexibility is paramount; AM is the perfect tool to deal with this, as the layered nature of additive manufacturing allows for almost infinite design flexibility during the design stage and allows easy part consolidation [4,60]. However, AM can limit modifications to the design in the field when compared with traditional modular design. From this perspective, it becomes clear that the use of AM in modular design requires the input of some non-modular design principles as well. Several of the basic modularization principles discussed in Section 2 are affected by this change in perspective, particularly standardization, serviceability, and customization (Table 2).

With these aspects in mind, both positive and negative, of the concept in mind, a reasonable definition of AM-enabled modular design (AMEMD) is:

> The process of decomposing a system into sets of functionalities or functions which can be performed by the design and additive manufacturing of multi-functional complex parts. These complex parts will be integrated together to produce the whole system. The functionality of the new system must be equal or superior to a system built from sub-assemblies of off-the-shelf parts, while containing fewer physical parts and interfaces.

Specific characteristics of this design methodology are:

- **Complexity vs standardization**: An increase in geometric complexity that results in fewer, but more function-dense parts is preferred over standardization of the parts and the use of assemblies of simple parts
- **Optimization**: Optimization for performance of complex parts is preferred over standardization
- **Design and integration**: Design teams for each of the complex parts focus on the design optimization and integration of the part into the system or product
- **Design prototyping**: System design prototypes can be produced quickly using AM technologies, aiding design and integration efforts
- **On-demand manufacturing**: For relatively simple systems, all AM parts, both original and replacement or upgrade parts, are manufactured on-demand directly from CAD data and are not kept in stock. For parts that belong to systems with certification requirements or that require constant stand-by spares, batches of parts can be made and kept as spares, with the option to make more on-demand if needed.
- **Serviceability**: While the parts are usually non-serviceable, they are easily replaced or upgraded, as long as the system allows access to the parts after completion (i.e., a "line-replaceable unit")
- **Part disposal**: Old parts that are worn out or upgraded are easily recycled due to the low-hazard status of most AM-processes materials
- **User customization**: Parts of the system are almost infinitely customizable and optimizable during the design, but usually not customizable in the field after production
- **Mass customization**: Several parts can be made at once using the same machine and the same batch of raw materials; there may be several of the same part or several different parts in a batch
- **Core mission**: There is only one core mission or set of missions for each system configuration, but this is offset by the increased speed and reduced cost of production and by the optimization of the design before use

Table 2. Modular principles directly affected by the combination-of-functions approach.

Standardization	AM allows the standardization of components, but it is usually more efficient and effective to prioritize optimization of the parts over standardization.
Serviceability	Most AM parts do not allow service on the parts after they are built and must simply be replaced when they wear out. The AM materials are typically highly recyclable and may even be recycled and made into new raw material by the part manufacturer, drastically reducing costs and simplifying the supply chain. The great reduction in cost, both from the elimination of tooling and from recycled materials, could offset the disadvantages of using non-serviceable parts.
User customization	AM-based design could potentially reduce the number of parts in the system, reducing the ability of the customer to modify the system. However, this can be offset by the ability to more completely customize the system during the design stage, reducing the need for the user to modify or optimize it in service. The need for a do-it-all system is greatly reduced when the system can be optimized for its mission at a low cost, as is the case with AM-enabled design.

4. Domain Applicability

The method presented is not be universally applicable to all types of products and systems and would require a defined domain of validity. The most important restrictions on the use of this design methodology are the physical size of parts, the available batch sizes, material requirements, and the designer's level of skill and experience with the technologies. In addition to these concerns, the designer must also ensure that the proper design-for-manufacturability constraints [29] are applied on the design to ensure that it is feasible for AM. Not all parts and systems will be best made using AM; this determination should be made early in the design process and the AMEMD only used if the complex parts can be feasibly made using AM. As discussed in the introduction, the practicality of using any DFAM method is dependent on the satisfaction of basic manufacturability constraints and the design considerations.

The three fundamental manufacturing aspects of all AM processes are the state of the raw material, the method of producing layers of raw material, and the process for fusing the layers together [17,18,27,29]. From these, the basic manufacturability constraints can be derived for a particular process; there are seven basic process, defined by these aspects, and each have their own set of constraints. Before any design can commence that may use AM, it should be ensured by the designer that the constraints can be met, since AM will not be a feasible manufacturing method otherwise and DFAM is not applicable.

For the combination-of-functions approach presented in this paper, the most important *design* consideration is the feasibility of consolidating or re-combining the basic parts or functions of the system. In cases where the design is new and the designer simply wishes to design several functions into one part, this is typically not a problem as the design can consider DFAM from conception. However, this method is likely most applicable to the re-design of systems which may benefit from the use of larger and more complex parts to replace subsystems. Therefore, the domain restriction on the design would be driven by the part consolidation. The primary issues that have come up in the literature [5,7,44,106,111,112] concerning part consolidation or re-combination are in the areas of repair, maintenance, and component upgrades to the system. As discussed in the previous section, the difficulty in these three areas is inherent in AM-made parts and is not unique to consolidated or re-combined parts. AM parts are typically not easily repairable, with the exception of some metal parts which can be welded or otherwise repaired using traditional manufacturing methods. Maintenance may or may not be a serious issue with AM parts, depending on the material being used and the function of the part; for example, some sintered metal parts subjected to wear can be left only partially dense and infused with oil to prevent the need for active maintenance. On the other hand, parts made from plastics, ceramics, and some metals cannot be maintained easily and so must usually be replaced when needed. One of the important advantages of using general modular design is the ability to upgrade sub-assemblies and individual parts on demand; this advantage does not typically exist at all for AM parts and this must be considered by designers.

In terms of the three main areas of concern for manufacturing constraints and three for part consolidation/re-combination, as well as the "practical" considerations described above, the potential drawbacks can be mitigated by careful planning and evaluation early in the design. Only from careful analysis of these constraints can the domain applicability be established for a particular design. Figure 6 shows some of the main decision variables that must be considered for this set of considerations. Under each are some of the possible aspects or requirements for the system that need to be identified or defined in order to best evaluate the benefits or drawbacks of DFAM; note that this is not an exhaustive list and other considerations may be needed for large and complex projects. The earlier in the design phase that DFAM can be considered, the better. Early adoption can serve to both drive the design or re-design to ensure feasibility with AM and to provide an early (and less costly) exit and adoption of another design method if it should become clear that the AM is not the best choice for manufacturing the system.

In cases where the use of AMEMD is applicable for part of the system or product but not all, a hybrid system could also certainly be developed, as shown in Figure 7. In practice, the hybrid system would be more applicable, as the system components or sub-assemblies could be made using the most feasible method for the design. Before using this or any other design methodology, the designer should carefully complete a manufacturability analysis to ensure that no dead-ends are encountered when completing the design. This is especially important when the product or system being designed will be used in a dynamic system or otherwise pose a risk to human life if failure occurs during use. This is an area that has not yet been explored well in the design literature and merits further consideration in future research in this area.

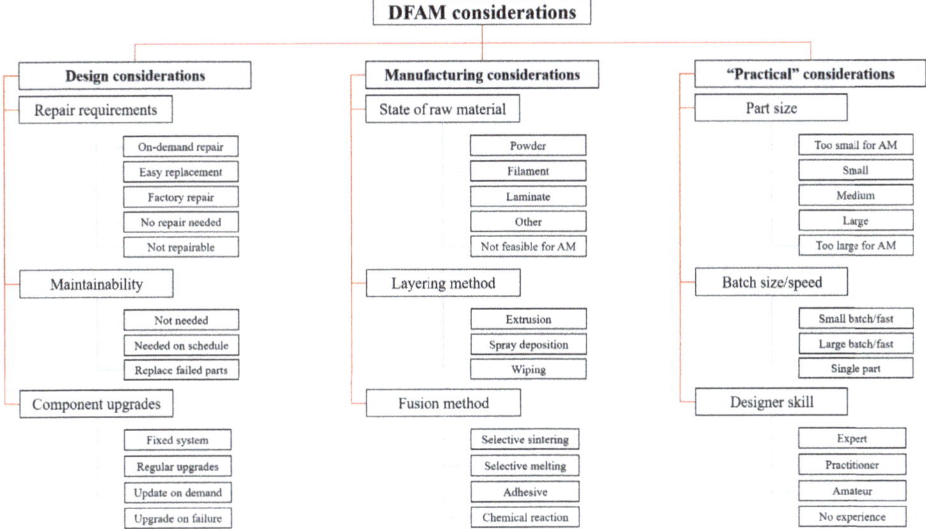

Figure 6. Some aspects to consider when evaluating the feasibility of DFAM.

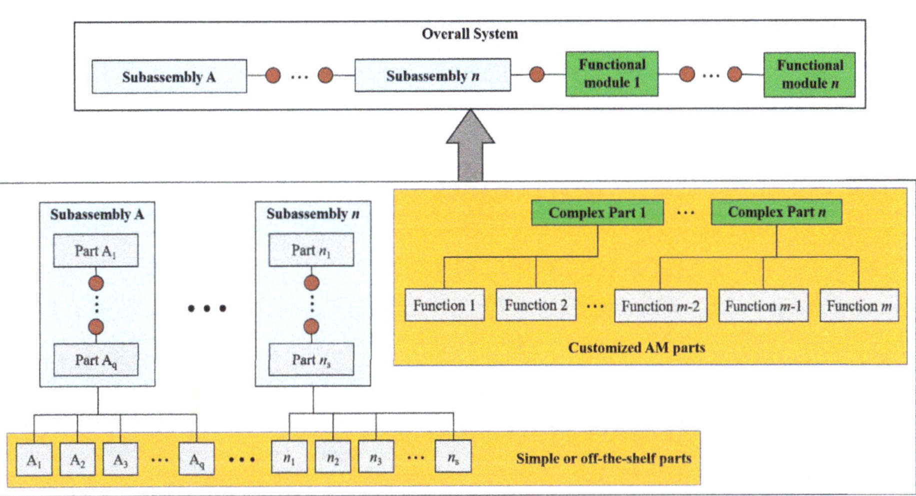

Figure 7. Hybrid subassembly-AMEMD system design concept.

5. Case Study: Re-Design of AIM-120 AMRAAM Airframe

5.1. Overview and Model Details

Presented in this section is a simple case study that was developed to demonstrate the procedure of re-design using AMEMD principles. In this case study, the system was taken and the parts combined into as few parts as possible in order to reduce their number and to reduce interfaces. The parts were then optimized on performance and mass to produce a system that is both optimized and has fewer components. One of the case studies presented by Hernandez et al. [60] was a fixed-fin model of an AIM-9X Sidewinder (Air Intercept Missile 9X). In the present work, another (more complex) missile airframe was studied, the AIM-120 AMRAAM (Advanced Medium-Range Air-to-Air Missile) (Figure 8); the airframe for this missile can be re-designed in order to decrease the number of parts while not fundamentally changing the external design. However, the internal design can be modified to utilize lattice structures or topology optimization that can be made using AM to optimize the design. Note that the exact design for this airframe is not available due to commercial and government protection, so the one presented here is a simplified model [113–116] but is as accurate as possible and will serve the purpose of this study.

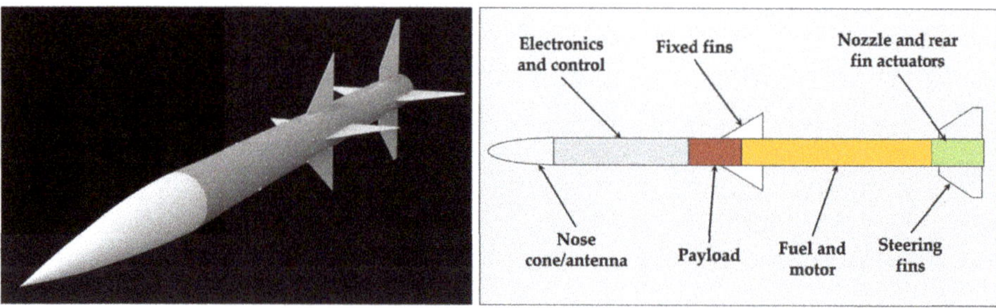

Figure 8. AIM-120 AMRAAM missile model [113,117]

Table 3 shows the basic list of components for this model, which consists of tube sections, control surfaces, and fasteners, for a total of 17 major parts and 60 fasteners; the exploded view of the airframe is shown in Figure 9. In this model, the airframe parts are assumed to be made from tough polymer materials, such as a phenolic, except for the fasteners and the rear control surfaces; those parts are assumed to be made from aluminum. The walls of the airframe are assumed to be 4 mm thick. The airframe will be subjected to some fatigue, but the duration of that fatigue will be limited to the few hours that the missile will be attached to an aircraft and be in flight after launch. Therefore, the anisotropic nature of the AM materials is not anticipated to be a threat to the integrity of the structure during storage or use. In use, certification will be required, as it will be for all AM-build flight hardware.

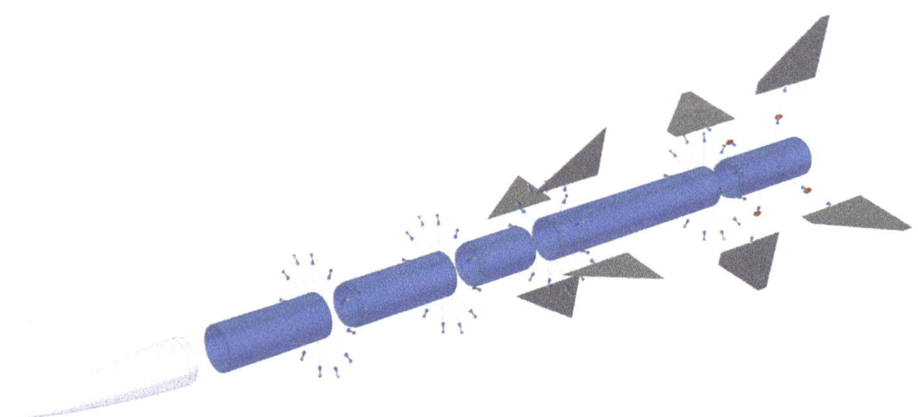

Figure 9. Exploded view: AIM-120 AMRAAM [113,117].

Table 3. AIM-120 AMRAAM airframe components.

Item	Component	Number in Model
1	Nose cone	1
2	Front electronics and control section	1
3	Rear electronics and control section	1
4	Payload housing section	1
5	Front control surfaces (fixed)	4
6	Fuel and motor housing section	1
7	Nozzle and rear control housing section	1
8	Rear control surfaces (adjustable)	4
9	Rear control surface actuator hinges	4
10	Fasteners	60

Both phenolic materials [118,119] and aluminum [120,121] are feasible to additively manufacture using the choice of one of several processes, and all of the parts are small enough to produce using existing AM processes. Therefore, this airframe can be made within the AM manufacturability constraints and therefore is a good candidate for AMEMD to improve its design and reduce the number of parts. In reviewing the basic design of this system, it becomes clear that a hybrid modular-AMEMD method would produce the most effective design since some of the sub-systems have moving parts and need to stay intact. Some of the sub-assemblies (Table 4) need to remain as such, but re-design can be accomplished on some other areas. The particular areas that should be examined are those where large parts can be combined or an interface between several parts needs to be used.

Table 4. AIM-120 airframe parts and sub-assemblies.

Sub-Assembly	Parts
Nose cone	Nose cone
Electronics housing	Front electronics and control section
Payload section	Rear electronics and control section Payload housing Front control surfaces
Fuel and motor housing	Fuel and motor housing section
Rear housing	Nozzle and rear control housing section Rear control surfaces Rear control surface actuator hinges

5.2. Examination of Part Combination Opportunities

Excluding the fasteners, there are 17 interfaces in the model as it is currently designed, each of which should be examined to find opportunities to combine parts into larger and more complex ones:

1. **Nose cone—upper electronics housing**: The nose cone holds the guidance for the missile and may need to be replaced or upgraded, so it is best to not eliminate this interface.
2. **Upper and lower electronics housing**: One of the details that can be derived from the promotional literature for the AIM-120 is that the upper section which holds the electronics is made in two pieces, which can be easily combined into a single piece, assuming that the internal components will allow this. It is assumed that it was originally made in two parts to allow customized connections to the internal parts; using AM to create an optimized internal structure to connect to the electronics should allow the combination of these two parts. In addition to removing an interface, 12 of the fasteners can also be eliminated. Preserving access to the electronics for assembly and maintenance would be an important design consideration during

the part re-combination. This should be checked to ensure that either no conflict exists or the AM part can be modified to account for it.

3. **Electronics housing to payload**: It is obvious that the payload section needs to be easily accessible to the user of the missile, so it would be best to retain this interface as-is.

4. **Payload to upper control surfaces**: For this missile, the upper control surfaces are fixed to the airframe and assumed to be made from the same material as the airframe. Therefore, the upper control surfaces can and should be integrated into the payload housing. The front fins could be combined with either the payload section or the motor/fuel section, but the best choice is the payload section; this is primarily to ensure that the air flows smoothly around the fins and that interface issues will not affect the leading surface of the control fin.

5. **Payload to motor/fuel section**: As previously stated, the payload section should be independent as much as possible, so this interface should remain. Access to the motor and fuel supply are also reasons to retain the interface between the payload and motor/fuel sections.

6. **Motor to nozzle section**: The rear control surfaces are driven by actuators near the nozzle and the nozzle and motor interface may need to be maintained and inspected.

7. **Nozzle section to rear control surface hinges**: Since the hinges need to be free-moving and connected to internal actuators, no combination can be accomplished.

8. **Hinges to rear control surfaces**: Since the rear control surfaces and hinges are both made from aluminum in this model, and their combination is relatively small, AMEMD makes sense for these parts.

5.3. Part Re-Combination and New System Model

After examination, three areas can benefit from the use of AMEMD; three sets of parts (the electronics housing, the mid-section fins and body, and the rear control surfaces) can be consolidated for a total reduction of parts from 17 to nine major components, plus the elimination of at least 12 fasteners. The upper and lower sections of the electronics housing (Figure 10a), the payload housing and its fins (Figure 10b), and the lower fins and their hinges (Figure 10c) can be recombined. In this case, the focus will be on reducing part count and optimizing the parts; future work will need to focus on other perspectives, such as ergonomics, safety, testing costs, and other such considerations.

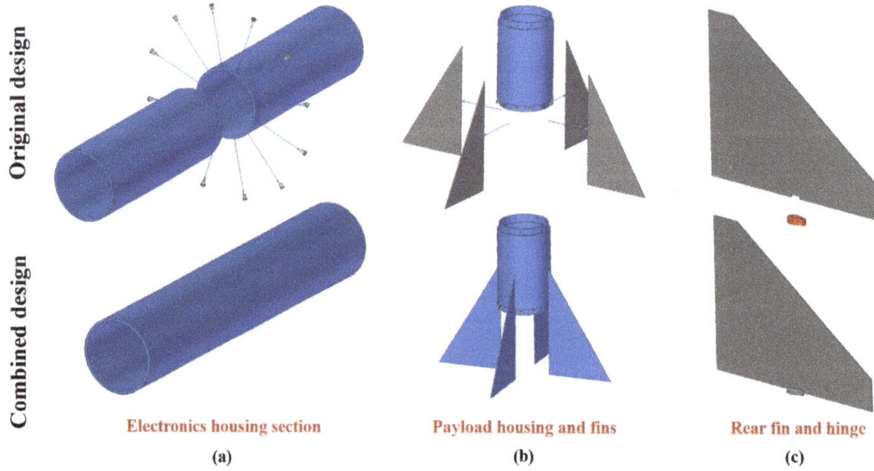

Figure 10. Part recombination opportunities for the AIM-120 AMRAAM airframe, (**a**) electronics housing; (**b**) payload section and upper fins, and (**c**) rear fins and actuator hinges.

The best design is certainly a hybrid of the AMEMD and the original design of the airframe, as shown in Figure 11. The nose cone and fuel/motor sections are left in their original design, the rear fins/hinges, the electronics housing, and the payload sections are combined to eliminate interfaces, and the rear housing section is a combination of both. Figure 12 shows the new airframe design, with the modified parts being highlighted in red; note the significant reduction of part count for the model with no fundamental change in the design.

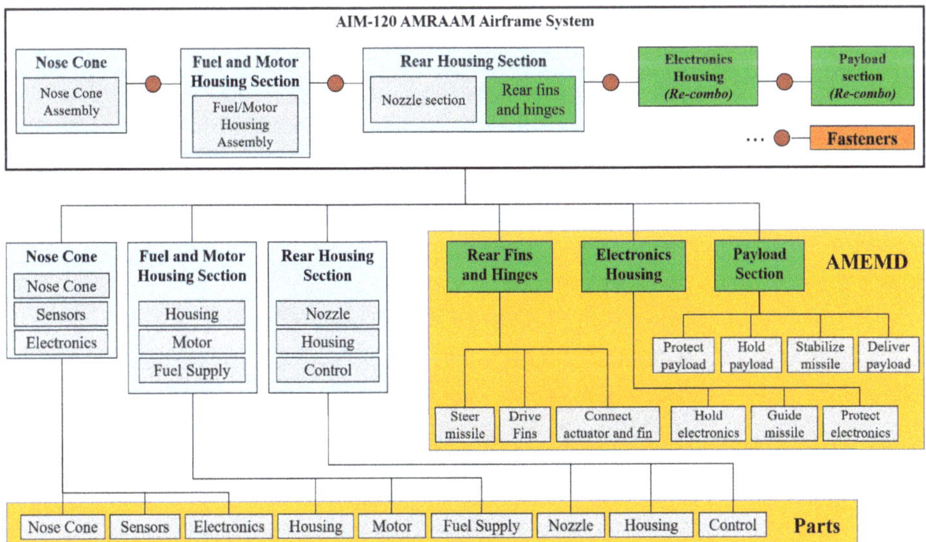

Figure 11. Proposed new AMRAAM airframe system design.

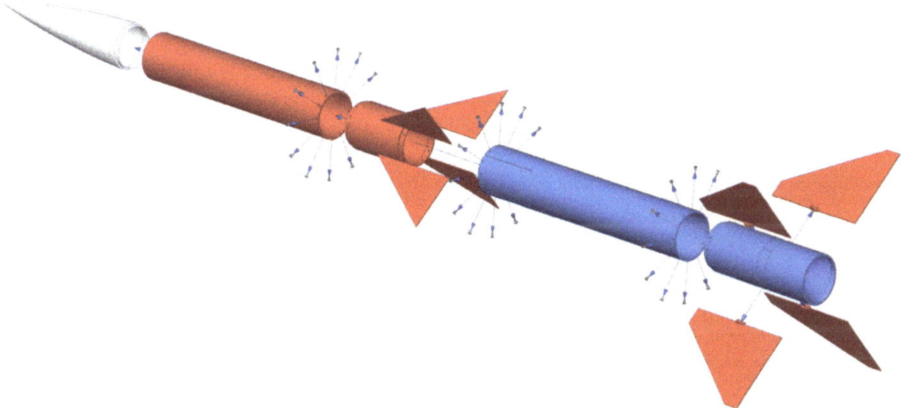

Figure 12. Proposed new AMRAAM airframe system design.

5.4. Part Re-Design and Optimization

In addition to combining functions into larger and more complex parts, the parts under study can be further optimized due to the design freedom offered by the use of AM to manufacture them.

5.4.1. Electronics Housing Optimization

For the new electronics housing, the external design and interfaces between the payload and the nose cone are fixed. However, internal modifications can be made using AM-enabled methods, particularly topology optimization and lattice design. In addition to the basic housing, mounting bosses and other internal structures can be integrated into the design. Figure 13 shows examples of this for the section under study, where the Pareto® TopOpt software was used to optimize the part. The basic housing (Figure 13a) can be redesigned with a box lattice structure, reducing the mass significantly (about 40% for this section) while maintaining structural integrity. If integrated internal features are needed, such as mounting bosses for the electronics, these can be integrated into the lattice as well; Figure 13b shows this with a body-centered cubic (BCC) lattice design. Both of these designs can easily be made using a variety of AM processes, subject only to the minimal length scale requirement for manufacturing [122–124]. In addition to the reduction of mass, the use of a lattice structure could also assist in cooling the electronics system.

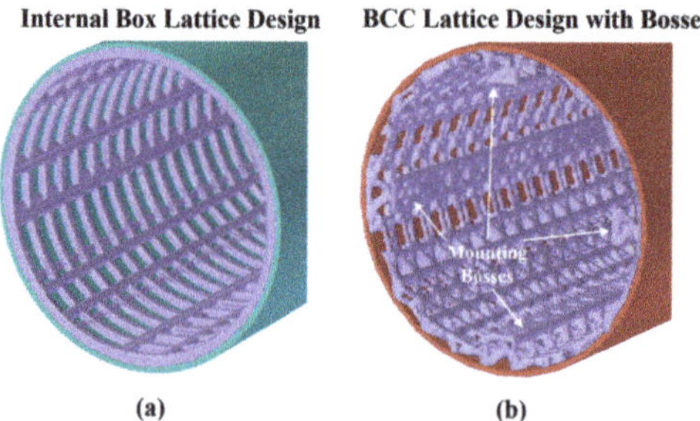

Figure 13. Internal lattice designs for electronics housing section where (**a**) is a simple lattice and (**b**) is a BCC lattice with integrated mounting bosses.

5.4.2. Payload and Upper Control Surfaces

In addition to the improvement of the internal structure, the payload section can also be integrated more fully with the control surfaces via a smoother transition. The smoother transition could reduce drag on the airframe, requiring less power to propel the missile; this, combined with the reduction in housing mass, increases the allowable payload mass and size without increasing the size of the missile or increasing fuel use. However, it is possible that other aspects could be negatively affected, such as frame integrity and stability during flight or the effect of a lighter airframe on the native sensor network in the missile.

Figure 14 shows an example solution for this, but many re-designs are feasible for this, determined primarily by the design and composition of the internal components. The fundamental purpose of the smoothing between the fins and the payload section is to reduce drag on the fins while also reinforcing them and preventing failure during flight. Since this missile is usually launched from a horizontal orientation, it is essential that the control surfaces are steady and reliable. The smoothing radius is a simple parameter that is easy to control during design and analysis, so two levels are shown. The radius can affect the streamlining of the missile, and therefore its stability and controllability during flight. Therefore, the designer should consider the best radius using both computational tools and physical testing.

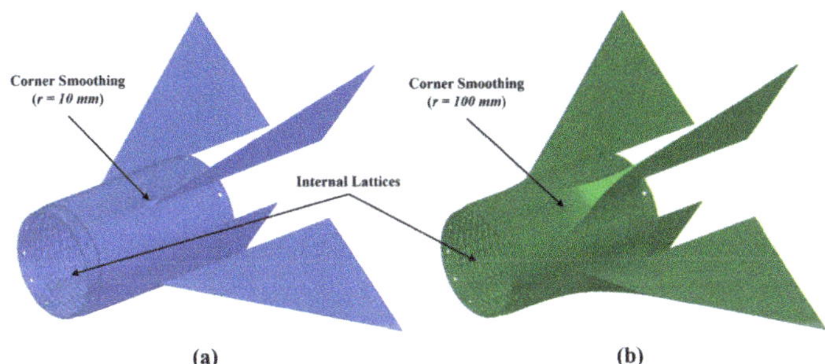

Figure 14. Redesign of payload section and upper control surfaces, where both new designs use an internal lattice and (**a**) uses a corner smoothing of $r = 10$ mm while (**b**) uses $r = 100$ mm.

5.4.3. Rear Control Surfaces

The re-design of the rear control surface is the least complex of the three new parts generated in this case study, requiring only the combination of the two parts and integrating. The combination of the two parts can be used as-is (Figure 15a) or the control surface can be re-designed to be more streamlined (Figure 15b), depending on the needs and preferences of the designer.

Figure 15. Redesign of rear control surface, (**a**) with a simple integration of the two functions and (**b**) with a re-design to streamline the airflow around the fin.

5.5. Design Evaluation

After re-design, the new system should be evaluated to ensure that the new design is at least equivalent to the old design; in theory, it should be superior to the old design but this must to be established in order to justify using AMEMD. Three major areas should be considered:

1. **Manufacturing**: The manufacturing process for the system should have been improved, with a particular focus on reducing production time and cost. In this case, a number of parts were re-combined which may reduce the assembly time; however, it is possible that the assembly time would increase if the condensed parts required any special tools or new processes to integrate with the other parts. For the airframe re-design, this may be an issue for the electronics housing if the internal components of the missile require easy access. The designer should consider this a trade-off when it is a problem; this may be mitigated during the design by producing a new assembly procedure which considers integration with the larger housing section.
2. **Functionality and Performance**: In order for AMEMD to be useful, it must produce some benefit to the designer. Therefore, it must be established that the new design gives at least one measurable benefit and that the general performance is at least as good as the old design. In this case, the performance and reliability of the missile must be at least as good as a missile which follows the original design. It was clear during the consolidation and optimization of the modified parts

that the weight was certainly reduced, which may have a positive impact on performance and control. An additional advantage which may be gained from using AM is the presence of a layered surface if untreated; this may reduce drag on the airframe by providing a small surface roughness to reduce the laminar fluid flow around the airframe.

3. **Usability**: In order for the new product or system to be useful, it must be reliable and useful for its core mission. The particular considerations for this and all AM-related designs are those discussed for the domain applicability, namely repairability, maintainability, and the possibility of upgrading the system. No matter how reliable the system is, there will eventually be breakdowns and needed part upgrades and so this must be considered during design. In terms of AM, repairability is sometimes possible but it should not be assumed for most AM parts. Maintainability and component upgrades are dependent on how the system is broken down; in general, any broken or worn parts made using AM would need to simply be replaced instead of being maintained or upgrades. However, this can be mitigated by the ability of AM to build complex and custom geometry and tailored materials, as this can reduce the need for maintenance. The on-demand manufacturability of AM parts would also reduce the need for part upgrades, as new part designs can be built as needed and the old one thrown away or recycled. For the missile airframe, the basic design is fairly well established, so it should be easy to design and manufacture new parts as needed in the field. Since the usability of the design is already established any upgrades or repairs would use a similar design to the old one, with similar reliability and usability.

6. Summary and Conclusions

The purpose of the present study was to explore the idea of using additive manufacturing to improve modular design of products by providing the means to use multi-function parts instead of sub-assemblies. Part consolidation and design modularity are often seen as conflicting design methods, but in the context of DFAM, they can be complementary for many systems. The elimination of the interfaces within the system is predicted to increase the reliability of the system, as well as simplifying it. The concept presented follows most of the basic principles of modular product design, as reviewed and discussed extensively, except for the principles of standardization, serviceability, and in-the-field customization. However, the loss of these capabilities within the modular design paradigm may not be a problem due to the design freedom from AM; standardization is no longer any more efficient than customization and serviceability and field customization are not major concerns, as new parts can be made on-demand in a few hours or days. This is, of course, subject to testing and certification requirements for any manufactured flight hardware made using AM. All AM processes have fairly restrictive manufacturability constraints, which must be considered carefully during design. In addition to the standard manufacturability constraints, the domain of applicability for this method is restricted by the size of the combined parts, the use of AM-processable materials, and on the skill and expertise of the designer to ensure manufacturability in the design. AM technologies are not efficient or preferable in all cases, so a study should be done before attempting such a design technique to ensure that it is feasible. The option to off-ramp from using an AM-based method should be built in as well, in case the design process proved non-feasible or too expensive for the problem as hand. The three most important material considerations are the choice of materials, the presence of fatigue stresses in the system (due to the anisotropic nature of the materials), and the smallest feature in the built part.

A broad case study was presented to demonstrate the concepts and to provide a practical example of applicability. A missile airframe was re-designed in a hybrid function-combination and traditional modular design, which resulted in the reduction of almost 50% of the parts and elimination of a number of fasteners. Optimization of the parts via AM-enabled digital means was demonstrated as well with the lattice design and streamlining of the parts.

This method is potentially very valuable for some systems, but should be considered one of several feasible methods until established as preferable and feasible by the designer. An important future research area for this concept is the certification of the resulting parts for use on actual systems. The case study presented in this study will require flight-hardware certification before use, but not all systems will require such a detailed and expensive process before use. For most simple systems and many others, this method may provide a more efficient design or re-design process which takes advantage of both the strengths of modularization and the design freedom of AM.

Author Contributions: C.C. and A.E.P. conceived and designed the study, C.C. preformed the modular design review, A.E.P. created the figures and case study, and A.E.P. and K.A.C. did the AM review and outline. C.L.C. and K.A.C. provided significant technical guidance and feedback on the ideas. All authors contributed to writing and polishing the manuscript for publication.

Funding: This research received no external funding

Acknowledgments: The authors would like to thank the following people for providing discussion and support for this work: James T. Allison, Dilcu Barnes, Paul D. Collopy, Kai James. Thanks to Kavian Niazi for granting permission to use the AIM-120 AMRAAM model from GrabCAD for the case study. The authors also thank the journal referees for their valuable feedback.

Conflicts of Interest: The authors declare no conflict of interest. No external funding was used to perform the work described in this study. Opinions and conclusions presented in this work are solely those of the authors.

References

1. Rosen, D.W. Computer-Aided Design for Additive Manufacturing of Cellular Structures. *Comput.-Aided Des. Appl.* **2007**, *4*, 585–594. [CrossRef]
2. Parthasarathy, J.; Starly, B.; Raman, S. A design for the additive manufacture of functionally graded porous structures with tailored mechanical properties for biomedical applications. *J. Manuf. Process.* **2011**, *13*, 160–170. [CrossRef]
3. Doubrovski, Z.; Verlinden, J.C.; Geraedts, J.M.P. Optimal Design for Additive Manufacturing: Opportunities and Challenges. In *Volume 9: 23rd International Conference on Design Theory and Methodology; 16th Design for Manufacturing and the Life Cycle Conference*; ASME: New York City, NY, USA, 2011. [CrossRef]
4. Yang, S.; Zhao, Y.F. Additive manufacturing-enabled design theory and methodology: A critical review. *Int. J. Adv. Manuf. Technol.* **2015**, *80*, 327–342. [CrossRef]
5. Schmelzle, J.; Kline, E.V.; Dickman, C.J.; Reutzel, E.W.; Jones, G.; Simpson, T.W. (Re)Designing for Part Consolidation: Understanding the Challenges of Metal Additive Manufacturing. *J. Mech. Des.* **2015**, *137*, 111404. [CrossRef]
6. Yang, S.; Santoro, F.; Sulthan, M.A.; Zhao, Y.F. A numerical-based part consolidation candidate detection approach with modularization considerations. *Res. Eng. Des.* **2018**. [CrossRef]
7. Oh, Y.; Zhou, C.; Behdad, S. Part decomposition and assembly-based (Re) design for additive manufacturing: A review. *Addit. Manuf.* **2018**, *22*, 230–242. [CrossRef]
8. Yang, S.; Tang, Y.; Zhao, Y.F. A new part consolidation method to embrace the design freedom of additive manufacturing. *J. Manuf. Process.* **2015**, *20*, 444–449. [CrossRef]
9. Jayal, A.; Badurdeen, F.; Dillon, O.; Jawahir, I. Sustainable manufacturing: Modeling and optimization challenges at the product, process and system levels. *CIRP J. Manuf. Sci. Technol.* **2010**, *2*, 144–152. [CrossRef]
10. Susman, G. *Integrating Design and Manufacturing for Competitive Advantage*; Oxford University Press: Oxford, UK, 1992.
11. Bralla, J.G. *Design for Manufacturability Handbook*; McGraw-Hill Education: New York City, NY, USA, 1998.
12. Gu, D.D.; Meiners, W.; Wissenbach, K.; Poprawe, R. Laser additive manufacturing of metallic components: Materials, processes and mechanisms. *Int. Mater. Rev.* **2012**, *57*, 133–164. [CrossRef]
13. Chan, K.S.; Koike, M.; Mason, R.L.; Okabe, T. Fatigue Life of Titanium Alloys Fabricated by Additive Layer Manufacturing Techniques for Dental Implants. *Metall. Mater. Trans. A* **2012**, *44*, 1010–1022. [CrossRef]
14. Jardini, A.L.; Larosa, M.A.; de Carvalho Zavaglia, C.A.; Bernardes, L.F.; Lambert, C.S.; Kharmandayan, P.; Calderoni, D.; Filho, R.M. Customised titanium implant fabricated in additive manufacturing for craniomaxillofacial surgery. *Virtual Phys. Prototyp.* **2014**, *9*, 115–125. [CrossRef]
15. Salmi, M.; Tuomi, J.; Paloheimo, K.S.; Björkstrand, R.; Paloheimo, M.; Salo, J.; Kontio, R.; Mesimäki, K.; Mäkitie, A.A. Patient-specific reconstruction with 3D modeling and DMLS additive manufacturing. *Rapid Prototyp. J.* **2012**, *18*, 209–214. [CrossRef]

16. Akmal, J.; Salmi, M.; Mäkitie, A.; Björkstrand, R.; Partanen, J. Implementation of Industrial Additive Manufacturing: Intelligent Implants and Drug Delivery Systems. *J. Funct. Biomater.* **2018**, *9*, 41. [CrossRef] [PubMed]

17. Guo, N.; Leu, M.C. Additive manufacturing: Technology, applications and research needs. *Front. Mech. Eng.* **2013**, *8*, 215–243. [CrossRef]

18. Gibson, I.; Rosen, D.; Stucker, B. *Additive Manufacturing Technologies: 3D Printing, Rapid Prototyping, and Direct Digital Manufacturing*; Springer: Berlin, Germany, 2016.

19. Lei, N.; Yao, X.; Moon, S.K.; Bi, G. An additive manufacturing process model for product family design. *J. Eng. Des.* **2016**, *27*, 751–767. [CrossRef]

20. Agard, B.; Bassetto, S. Modular design of product families for quality and cost. *Int. J. Prod. Res.* **2013**, *51*, 1648–1667. [CrossRef]

21. Jiao, J.R.; Simpson, T.W.; Siddique, Z. Product family design and platform-based product development: A state-of-the-art review. *J. Intell. Manuf.* **2007**, *18*, 5–29. [CrossRef]

22. Nayak, R.U.; Chen, W.; Simpson, T.W. A Variation-Based Method for Product Family Design. *Eng. Optim.* **2002**, *34*, 65–81. [CrossRef]

23. Gonzalez-Zugasti, J.P.; Otto, K.N.; Baker, J.D. Assessing value in platformed product family design. *Res. Eng. Des.* **2001**, *13*, 30–41. [CrossRef]

24. Patterson, A.E.; Thomas, L.D.; Messimer, S.L. Disruptive Effects of Additive Technologies on SE Product Lifecycle. In Proceedings of the JANNAF Technical Interchange Conference, Huntsville, AL, USA, 23–25 August 2016. [CrossRef]

25. Liu, P.; Huang, S.H.; Mokasdar, A.; Zhou, H.; Hou, L. The impact of additive manufacturing in the aircraft spare parts supply chain: Supply chain operation reference (SCOR) model based analysis. *Prod. Plan. Control* **2013**, *25*, 1169–1181. [CrossRef]

26. Khajavi, S.H.; Partanen, J.; Holmström, J. Additive manufacturing in the spare parts supply chain. *Comput. Ind.* **2014**, *65*, 50–63. [CrossRef]

27. ASTM. F2792-12a: Standard Terminology for Additive Manufacturing Technologies. *ASTM Stand.* **2012**. [CrossRef]

28. Patterson, A.E.; Collopy, P.; Messimer, S.L. *State-of-the-Art Survey of Additive Manufacturing Technologies, Methods, and Materials*; Technical Report, Report Number UAH 2015-06; Department of Industrial & Systems Engineering and Engineering Management, University of Alabama in Huntsville: Huntsville, IL, USA, 2015. [CrossRef]

29. Patterson, A.E.; Allison, J.T. Manufacturability Constraint Formulation for Design Under Hybrid Additive-Subtractive Manufacturing. In *ASME 2018 IDETC/CIE: 23rd Deign for Manufacturing and the Life Cycle Conference (DFMLC)*; ASME: Quebec City, QC, Canada, 2018. [CrossRef]

30. Murthy, D.; Østerås, T.; Rausand, M. Component reliability specification. *Reliab. Eng. Syst. Saf.* **2009**, *94*, 1609–1617. [CrossRef]

31. Aal, A.; Polte, T. On component reliability and system reliability for automotive applications. In Proceedings of the IEEE International Integrated Reliability Workshop Final Report, South Lake Tahoe, CA, USA, 14–18 October 2012. [CrossRef]

32. Elsayed, E.A. *Reliability Engineering (Wiley Series in Systems Engineering and Management)*; Wiley: Hoboken, NJ, USA, 2012.

33. Hohenbichler, M.; Rackwitz, R. First-order concepts in system reliability. *Struct. Saf.* **1982**, *1*, 177–188. [CrossRef]

34. Chern, M.S. On the computational complexity of reliability redundancy allocation in a series system. *Oper. Res. Lett.* **1992**, *11*, 309–315. [CrossRef]

35. Frey, D.; Palladino, J.; Sullivan, J.; Atherton, M. Part count and design of robust systems. *Syst. Eng.* **2007**, *10*, 203–221. [CrossRef]

36. ElMaraghy, W.; ElMaraghy, H.; Tomiyama, T.; Monostori, L. Complexity in engineering design and manufacturing. *CIRP Ann.* **2012**, *61*, 793–814. [CrossRef]

37. Xu, Z.; Kuo, W.; Lin, H.H. Optimization limits in improving system reliability. *IEEE Trans. Reliab.* **1990**, *39*, 51–60. [CrossRef]

38. Billinton, R.; Allan, R.N. *Reliability Evaluation of Engineering Systems*; Springer: Berlin, Germany, 1983. [CrossRef]

39. Anderson, K.B.; Lockwood, S.Y.; Martin, R.S.; Spence, D.M. A 3D Printed Fluidic Device that Enables Integrated Features. *Anal. Chem.* **2013**, *85*, 5622–5626. [CrossRef] [PubMed]

40. Dind, A.; Lufkin, S.; Rey, E. A Modular Timber Construction System for the Sustainable Vertical Extension of Office Buildings. *Designs* **2018**, *2*, 30. [CrossRef]

41. Sharafi, P.; Nemati, S.; Samali, B.; Ghodrat, M. Development of an Innovative Modular Foam-Filled Panelized System for Rapidly Assembled Postdisaster Housing. *Buildings* **2018**, *8*, 97. [CrossRef]

42. Zhang, X.; Huang, Y.; Li, L.; Yeh, W.C. Power and Capacity Consensus Tracking of Distributed Battery Storage Systems in Modular Microgrids. *Energies* **2018**, *11*, 1439. [CrossRef]

43. Mishra, A.; Mondini, A.; Dottore, E.D.; Sadeghi, A.; Tramacere, F.; Mazzolai, B. Modular Continuum Manipulator: Analysis and Characterization of Its Basic Module. *Biomimetics* **2018**, *3*, 3. [CrossRef]

44. Demir, İ.; Aliaga, D.G.; Benes, B. Near-convex decomposition and layering for efficient 3D printing. *Addit. Manuf.* **2018**, *21*, 383–394. [CrossRef]

45. Liu, J. Guidelines for AM part consolidation. *Virtual Phys. Prototyp.* **2016**, *11*, 133–141. 2016.1175154. [CrossRef]

46. Yang, S.; Zhao, Y.F. Additive Manufacturing-Enabled Part Count Reduction: A Lifecycle Perspective. *J. Mech. Des.* **2018**, *140*, 031702. [CrossRef]

47. Yang, S.; Zhao, Y. Conceptual design for assembly in the context of additive manufacturing. In Proceedings of the 26th Annual International Solid Freeform Fabrication Symposium, Austin, TX, USA, 10–12 August 2016.

48. Thompson, M.K.; Moroni, G.; Vaneker, T.; Fadel, G.; Campbell, R.I.; Gibson, I.; Bernard, A.; Schulz, J.; Graf, P.; Ahuja, B.; et al. Design for Additive Manufacturing: Trends, opportunities, considerations, and constraints. *CIRP Ann.* **2016**, *65*, 737–760. [CrossRef]

49. Mhapsekar, K.; McConaha, M.; Anand, S. Additive Manufacturing Constraints in Topology Optimization for Improved Manufacturability. *J. Manuf. Sci. Eng.* **2018**, *140*, 051017. [CrossRef]

50. Meisel, N.; Williams, C. An Investigation of Key Design for Additive Manufacturing Constraints in Multimaterial Three-Dimensional Printing. *J. Mech. Des.* **2015**, *137*, 111406. [CrossRef]

51. Kok, Y.; Tan, X.; Wang, P.; Nai, M.; Loh, N.; Liu, E.; Tor, S. Anisotropy and heterogeneity of microstructure and mechanical properties in metal additive manufacturing: A critical review. *Mater. Des.* **2018**, *139*, 565–586. [CrossRef]

52. Popovich, A.A.; Sufiiarov, V.S.; Borisov, E.V.; Polozov, I.A.; Masaylo, D.V.; Grigoriev, A.V. Anisotropy of mechanical properties of products manufactured using selective laser melting of powdered materials. *Russ. J. Non-Ferr. Met.* **2017**, *58*, 389–395. [CrossRef]

53. Ahn, S.H.; Montero, M.; Odell, D.; Roundy, S.; Wright, P.K. Anisotropic material properties of fused deposition modeling ABS. *Rapid Prototyp. J.* **2002**, *8*, 248–257. [CrossRef]

54. Kaji, F.; Barari, A. Evaluation of the Surface Roughness of Additive Manufacturing Parts Based on the Modelling of Cusp Geometry. *IFAC-PapersOnLine* **2015**, *48*, 658–663. [CrossRef]

55. Townsend, A.; Senin, N.; Blunt, L.; Leach, R.; Taylor, J. Surface texture metrology for metal additive manufacturing: A review. *Precis. Eng.* **2016**, *46*, 34–47. [CrossRef]

56. Alfieri, V.; Argenio, P.; Caiazzo, F.; Sergi, V. Reduction of Surface Roughness by Means of Laser Processing over Additive Manufacturing Metal Parts. *Materials* **2016**, *10*, 30. [CrossRef] [PubMed]

57. Patterson, A.E.; Messimer, S.L.; Farrington, P.A. Overhanging Features and the SLM/DMLS Residual Stresses Problem: Review and Future Research Need. *Technologies* **2017**, *5*, 15. [CrossRef]

58. Mukherjee, T.; Zhang, W.; DebRoy, T. An improved prediction of residual stresses and distortion in additive manufacturing. *Comput.l Mater. Sci.* **2017**, *126*, 360–372. [CrossRef]

59. Li, C.; Liu, Z.; Fang, X.; Guo, Y. Residual Stress in Metal Additive Manufacturing. *Procedia CIRP* **2018**, *71*, 348–353. [CrossRef]

60. Hernandez, R.; Singh, H.; Messimer, S.; Patterson, A. Design and Performance of Modular 3-D Printed Solid-Propellant Rocket Airframes. *Aerospace* **2017**, *4*, 17. [CrossRef]

61. Chung, W.H.; Kremer, G.E.O.; Wysk, R.A. Life cycle implications of product modular architectures in closed-loop supply chains. *Int. J. Adv. Manuf. Technol.* **2013**, *70*, 2013–2028. [CrossRef]

62. Gershenson, J.K.; Prasad, G.J.; Zhang, Y. Product modularity: Definitions and benefits. *J. Eng. Des.* **2003**, *14*, 295–313. [CrossRef]

63. Baldwin, C.; Clark, K. Managing in the Age of Modularity. *Harvard Bus. Rev.* **1997**, *75*, 84–93.

64. Bonvoisin, J.; Halstenberg, F.; Buchert, T.; Stark, R. A systematic literature review on modular product design. *J. Eng. Des.* **2016**, *27*, 488–514. [CrossRef]
65. Todorova, G.; Burisin, B. Do firms innovate through integrated modular designs? Disentangling the effects of types of modularity on types of innovation. In *KITeS Knowledge, Internationalization, and Technology Studies*; Technical Report; Bocconi University: Milan, Italy, 2009.
66. Ulrich, K. Fundamentals of Product Modularity. In *Management of Design*; Springer: Berlin, Germany, 1994; pp. 219–231._12. [CrossRef]
67. Ozman, M. Modularity, Industry Life Cycle and Open Innovation. *J. Technol. Manag. Innovat.* **2011**, *6*, 26–34. [CrossRef]
68. Ettlie, J.E.; Kubarek, M. Design Reuse in Manufacturing and Services. *J. Prod. Innovat. Manag.* **2008**, *25*, 457–472. [CrossRef]
69. Kassi, T.; Leisti, S.; Puheloinen, T. Impact of product modular design on agile manufacturing. *Mechanika* **2008**, *74*, 56–62.
70. Ericsson, A.; Erixon, G. *Controlling Design Variants: Modular Product Platforms*; Society of Manufacturing Engineers: Dearborn, MI, USA, 1999.
71. Kreng, V.B.; Lee, T.P. QFD-based modular product design with linear integer programming—A case study. *J. Eng. Des.* **2004**, *15*, 261–284. [CrossRef]
72. Kremer, G.E.O.; Gupta, S. Analysis of modularity implementation methods from an assembly and variety viewpoints. *Int. J. Adv. Manuf. Technol.* **2012**, *66*, 1959–1976. [CrossRef]
73. Okudan, G.E.; Chiu, M.C.; Kim, T.H. Perceived feature utility-based product family design: A mobile phone case study. *J. Intell. Manuf.* **2012**, *24*, 935–949. [CrossRef]
74. Agard, B.; Kusiak, A. Data-mining-based methodology for the design of product families. *Int. J. Prod. Res.* **2004**, *42*, 2955–2969. [CrossRef]
75. Otto, K.N.; Wood, K.L. Product Evolution: A Reverse Engineering and Redesign Methodology. *Res. Eng. Des.* **1998**, *10*, 226–243. [CrossRef]
76. Stone, R.B.; McAdams, D.A.; Kayyalethekkel, V.J. A product architecture-based conceptual DFA technique. *Des. Stud.* **2004**, *25*, 301–325. [CrossRef]
77. Asan, U.; Polat, S.; Serdar, S. An integrated method for designing modular products. *J. Manuf. Technol. Manag.* **2004**, *15*, 29–49. [CrossRef]
78. Wasley, N.S.; Lewis, P.K.; Mattson, C.A.; Ottosson, H.J. Experimenting with concepts from modular product design and multi-objective optimization to benefit people living in poverty. *Dev. Eng.* **2017**, *2*, 29–37. [CrossRef]
79. Yigit, A.S.; Ulsoy, A.G.; Allahverdi, A. Optimizing modular product design for reconfigurable manufacturing. *J. Intell. Manuf.* **2002**, *13*, 309–316. [CrossRef]
80. Levin, M.S. Combinatorial optimization in system configuration design. *Autom. Remote Control* **2009**, *70*, 519–561. [CrossRef]
81. Kuo, T.C.; Huang, S.H.; Zhang, H.C. Design for manufacture and design for 'X': Concepts, applications, and perspectives. *Comput. Ind. Eng.* **2001**, *41*, 241–260. [CrossRef]
82. Gatenby, D.A.; Foo, G. Design for X (DFX): Key to Competitive, Profitable Products. *AT&T Tech. J.* **1990**, *69*, 2–13. [CrossRef]
83. Chiu, M.C.; Okudan, G. An Integrative Methodology for Product and Supply Chain Design Decisions at the Product Design Stage. *J. Mech. Des.* **2011**, *133*, 021008. [CrossRef]
84. Gu, P.; Hashemian, M.; Sosale, S.; Rivin, E. An Integrated Modular Design Methodology for Life-Cycle Engineering. *CIRP Ann.* **1997**, *46*, 71–74. [CrossRef]
85. Ma, J.; Kremer, G.E.O. A sustainable modular product design approach with key components and uncertain end-of-life strategy consideration. *Int. J. Adv. Manuf. Technol.* **2015**, *85*, 741–763. [CrossRef]
86. Mutingi, M.; Dube, P.; Mbohwa, C. A Modular Product Design Approach for Sustainable Manufacturing in A Fuzzy Environment. *Procedia Manuf.* **2017**, *8*, 471–478. [CrossRef]
87. Kristianto, Y.; Helo, P. Product architecture modularity implications for operations economy of green supply chains. *Transp. Res. Part E Logist. Transp. Rev.* **2014**, *70*, 128–145. [CrossRef]
88. Yang, Q.; Yu, S.; Sekhari, A. A modular eco-design method for life cycle engineering based on redesign risk control. *Int. J. Adv. Manuf. Technol.* **2011**, *56*, 1215–1233. [CrossRef]

89. Wang, Q.; Tang, D.; Yin, L.; Yang, J. A Method for Green Modular Design Considering Product Platform Planning Strategy. *Procedia CIRP* **2016**, *56*, 40–45. [CrossRef]

90. Booker, J.; Swift, K.; Brown, N. Designing for assembly quality: Strategies, guidelines and techniques. *J. Eng. Des.* **2005**, *16*, 279–295. [CrossRef]

91. Gu, P.; Sosale, S. Product modularization for life cycle engineering. *Robot. Comput.-Integr. Manuf.* **1999**, *15*, 387–401. [CrossRef]

92. Sosa, M.E.; Eppinger, S.D.; Rowles, C.M. Identifying Modular and Integrative Systems and Their Impact on Design Team Interactions. *J. Mech. Des.* **2003**, *125*, 240. [CrossRef]

93. Li, Z.S.; Mobin, M.S. System reliability assessment incorporating interface and function failure. In Proceedings of the Annual Reliability and Maintainability Symposium (RAMS), Palm Harbor, FL, USA, 26–29 January 2015. [CrossRef]

94. Ebeling, C.E. *An Introduction to Reliability and Maintainability Engineering*; Waveland Press: Long Grove, IL, USA, 2009.

95. Blanchard, B.S.; Fabrycky, W.J. *Systems Engineering and Analysis*, 4th ed.; Prentice Hall: Upper Saddle River, NJ, USA, 2005.

96. NASA. *NASA Systems Engineering Handbook: NASA/Sp-2016-6105 Rev2 - Full Color Version*; 12th Media Services: Suwanee, GA, USA, 2017.

97. Sommerer, S.; Guevara, M.D.; Landis, M.A.; Rizzuto, J.M.; Sheppard, J.M.; Grant, C.J. System-of-Systems Engineering in Air and Missile Defense. *Johns Hopkins APL Tech. Dig.* **2012**, *31*, 5–20.

98. Becker, R.; Grzesiak, A.; Henning, A. Rethink assembly design. *Assem. Autom.* **2005**, *25*, 262–266. [CrossRef]

99. Salemi, B.; Moll, M.; Shen, W.-M. SUPERBOT: A Deployable, Multi-Functional, and Modular Self-Reconfigurable Robotic System. In Proceedings of the IEEE/RSJ International Conference on Intelligent Robots and Systems, Beijing, China, 9–15 October 2006. [CrossRef]

100. Kelmar, L.; Khosla, P. Automatic generation of kinematics for a reconfigurable modular manipulator system. In Proceedings of the IEEE International Conference on Robotics and Automation, Philadelphia, PA, USA, 24–29 April 1988. [CrossRef]

101. Berman, B. 3-D printing: The new industrial revolution. *Bus. Horiz.* **2012**, *55*, 155–162. [CrossRef]

102. Petrick, I.J.; Simpson, T.W. 3D Printing Disrupts Manufacturing: How Economies of One Create New Rules of Competition. *Res.-Technol. Manag.* **2013**, *56*, 12–16. [CrossRef]

103. Thomas, D. Costs, benefits, and adoption of additive manufacturing: A supply chain perspective. *Int. J. Adv. Manuf. Technol.* **2015**, *85*, 1857–1876. [CrossRef] [PubMed]

104. Charter, M. Sustainable Product Design. In *The Durable Use of Consumer Products*; Springer: Berlin, Germany, 1998; pp. 57–68._5. [CrossRef]

105. Allison, N.; Richards, J. Current status and future trends for disposable technology in the biopharmaceutical industry. *J. Chem. Technol. Biotechnol.* **2013**, *89*, 1283–1287. [CrossRef]

106. Maidin, S.B.; Campbell, I.; Pei, E. Development of a design feature database to support design for additive manufacturing. *Assem. Autom.* **2012**, *32*, 235–244. [CrossRef]

107. Boothroyd, G.; Knight, W. Design for assembly. *IEEE Spectrum.* **1993**, *30*, 53–55. [CrossRef]

108. Mok, H.S.; Kim, C.H.; Kim, C.B. Automation of mold designs with the reuse of standard parts. *Expert Syst. Appl.* **2011**, *38*, 12537–12547. [CrossRef]

109. Culley, S.J. Classification approaches for standard parts to aid design reuse. *Proc. Inst. Mech. Eng. Part B J. Eng. Manuf.* **1999**, *213*, 203–207. [CrossRef]

110. Novak, S.; Eppinger, S.D. Sourcing By Design: Product Complexity and the Supply Chain. *Manag. Sci.* **2001**, *47*, 189–204. [CrossRef]

111. Laverne, F.; Segonds, F.; Anwer, N.; Coq, M.L. Assembly Based Methods to Support Product Innovation in Design for Additive Manufacturing: An Exploratory Case Study. *J. Mech. Des.* **2015**, *137*, 121701. [CrossRef]

112. Tang, Y.; Yang, S.; Zhao, Y.F. Sustainable Design for Additive Manufacturing Through Functionality Integration and Part Consolidation. In *Handbook of Sustainability in Additive Manufacturing*; Springer: Singapore, 2016; pp. 101–144._6. [CrossRef]

113. Niazi, K. AIM-120 AMRAAM. GrabCAD Model. Available online: https://grabcad.com/library/aim-120-amraam-1 (accessed on 30 May 2018).

114. Gonzalez-Sanchez, M.; Amezquita-Brooks, L.; Liceaga-Castro, E.; del C Zambrano-Robledo, P. Simplifying quadrotor controllers by using simplified design models. In Proceedings of the 52nd IEEE Conference on Decision and Control, Florence, Italy, 10–13 December 2013. [CrossRef]

115. Kim, C.; Mijar, A.; Arora, J. Development of simplified models for design and optimization of automotive structures for crashworthiness. *Struct. Multidiscip. Optim.* **2001**, *22*, 307–321. [CrossRef]

116. Lu, X.; Xie, L.; Yu, C.; Lu, X. Development and application of a simplified model for the design of a super-tall mega-braced frame-core tube building. *Eng. Struct.* **2016**, *110*, 116–126. [CrossRef]

117. AMRAAM Missile: Modern, Versatile, and Proven. Available online: https://www.raytheon.com/capabilities/products/amraam (accessed on 15 June 2018).

118. Snelling, D.; Williams, C.; Druschitz, A. A comparison of binder burnout and mechanical characteristics of printed and chemically bonded sand molds. In Proceedings of the 2014 Solid Freeform Fabrication Symposium, Austin, TX, USA, 13–15 August 2014.

119. Kumar, S.; Kruth, J.P. Composites by rapid prototyping technology. *Mater. Des.* **2010**, *31*, 850–856. [CrossRef]

120. Louvis, E.; Fox, P.; Sutcliffe, C.J. Selective laser melting of aluminium components. *J. Mater. Process. Technol.* **2011**, *211*, 275–284. [CrossRef]

121. Aboulkhair, N.T.; Everitt, N.M.; Maskery, I.; Ashcroft, I.; Tuck, C. Selective laser melting of aluminum alloys. *MRS Bull.* **2017**, *42*, 311–319. [CrossRef]

122. Vyatskikh, A.; Delalande, S.; Kudo, A.; Zhang, X.; Portela, C.M.; Greer, J.R. Additive manufacturing of 3D nano-architected metals. *Nat. Commun.* **2018**, *9*. [CrossRef] [PubMed]

123. Chang, J.; He, J.; Mao, M.; Zhou, W.; Lei, Q.; Li, X.; Li, D.; Chua, C.K.; Zhao, X. Advanced Material Strategies for Next-Generation Additive Manufacturing. *Materials* **2018**, *11*, 166. [CrossRef] [PubMed]

124. Slotwinski, J.A.; Garboczi, E.J.; Hebenstreit, K.M. Porosity Measurements and Analysis for Metal Additive Manufacturing Process Control. *J. Res. Nat. Inst. Stand. Technol.* **2014**, *119*, 494. [CrossRef] [PubMed]

Article

Tetrahedron-Based Porous Scaffold Design for 3D Printing

Ye Guo [1,*], Ke Liu [2] and Zeyun Yu [1,*]

[1]   Department of Computer Science, University of Wisconsin—Milwaukee, Milwaukee, WI 53211, USA
[2]   Lattice Engines Inc., San Mateo, CA 94404, USA; lkhoho@gmail.com
*   Correspondence: yeguo@uwm.edu (Y.G.); yuz@uwm.edu (Z.Y.)

Received: 9 January 2019; Accepted: 4 February 2019; Published: 18 February 2019

Abstract: Tissue repairing has been the ultimate goal of surgery, especially with the emergence of reconstructive medicine. A large amount of research devoted to exploring innovative porous scaffold designs, including homogeneous and inhomogeneous ones, have been presented in the literature. The triply periodic minimal surface has been a versatile source of biomorphic structure design due to its smooth surface and high interconnectivity. Nonetheless, many 3D models are often rendered in the form of triangular meshes for its efficiency and convenience. The requirement of regular hexahedral meshes then becomes one of limitations of the triply periodic minimal surface method. In this paper, we make a successful attempt to generate microscopic pore structures using tetrahedral implicit surfaces. To replace the conventional Cartesian coordinates, a new coordinates system is built based on the perpendicular distances between a point and the tetrahedral faces to capture the periodicity of a tetrahedral implicit surface. Similarly to the triply periodic minimal surface, a variety of tetrahedral implicit surfaces, including P-, D-, and G-surfaces are defined by combinations of trigonometric functions. We further compare triply periodic minimal surfaces with tetrahedral implicit surfaces in terms of shape, porosity, and mean curvature to discuss the similarities and differences of the two surfaces. An example of femur scaffold construction is provided to demonstrate the detailed process of modeling porous architectures using the tetrahedral implicit surface.

Keywords: porous scaffold design; tetrahedral implicit surface modeling; triply periodic minimal surface; 3D printing

---

1. Introduction

Tissue Engineering (TE), as an interdisciplinary discipline, is addressed to create artificial, functional three-dimensional (3D) tissues and organs. The superb porous scaffold plays a critical role in cell proliferation and differentiation phases. There are a few basic, widely accepted criteria for building high-quality scaffolds. Firstly, the internal porous structure is required for interconnection and high porosity, and also has proper pore size. Secondly, the materials of scaffolds have to be biocompatible and bioresorbable with a controllable degradation rate. Thirdly, a certain level of mechanical strength is necessary to maintain the structure and support the weight from other parts of the body. Fourthly, a large surface area is of great importance to facilitate the delivery of nutrients and metabolic waste. While replication of the tissue's geometry is feasible when using advanced computer-aided design (CAD) and medical imaging techniques, the design of internal microstructures with the same bio-mechanical functionality as real tissue is more challenging.

A variety of fabrication methods have been proposed for scaffold manufacturing. Earlier on, Mikos et al. demonstrated a native scaffold design technique using salt leaching [1], while Mooney et al. proposed fabricating porous sponges of poly using gas-foaming [2]. Meanwhile, a few other conventional manufacturing methods, such as thermal-induced phase separation [3,4], porous ceramics [5], electrospinning [6,7], biomineralization [8,9], and phase-separation followed by freeze-drying [10] were also suggested for porous scaffold fabrication. Nevertheless, these methods have limitations with regard to accurately controlling pore morphologies and sizes. With the rapid development of additive manufacturing (AM) techniques and their applications in tissue engineering, a paradigm shift toward computer-aided design has occurred in tissue engineering. Common additive manufacturing (also known as 3D printing) technologies include fused deposition modeling (FDM), selective laser sintering (SLS), precision extrusion deposition (PED), and stereolithography (SLA) [11–14].

Various commercial CAD softwares, either based on constructive solid geometry (CSG) or boundary representation (B-Rep), have been commonly employed to assist in engineering design and modeling. Some widely used tools are CATIA (Dassault Systèmes, Vélizy-Villacoublay Cedex, France), NX (Siemens PLM Software, Plano, TX, USA), SolidWorks (Dassault Systèmes, Vélizy-Villacoublay, France), and Pro/Engineer (PTC, Needham, MA, USA) [15,16]. These CAD tools combine diverse standard solid primitives (cylinders, spheres, cubes, etc.) through Boolean operations to produce complicated 3D models. Since B-Rep methods describe the solid based on a combination of vertices, edges, and loops, a preliminary check is usually placed to ensure no gaps or overlaps among the boundaries of different cells [17]. In order to enrich the parametric library of scaffolds structures, a set of unit cells with bio-inspired features have also been revealed in some recent research [18,19]. Although the utilization of present CAD tools has largely reduced the time taken for scaffold design and automated the manufacturing process, the biomechanical properties of the resulting scaffolds are still relatively poor. As an alternative solution, a few image-based approaches that combine image processing and free-form fabrication techniques are also used for modeling 3D scaffold geometries. Due to their compatibility with real patient data, image-based methods can quickly create porous architectures by intersecting two 3D binary images, among which one depicts the outline of the defect and the other one consists of a stack of binary unit cells [14]. As an example, Hollister et al. defined the craniofacial scaffold's topology by specifically setting the voxels' densities within image design cubes [20]. Later, Smith et al. also used image-based design and CAD tools to generate an accurately sized and shaped scaffold for osseous tissue via selective laser sintering (SLS) [21]. Compared to other CAD scaffold design patterns, implicit surface modeling (ISM) methods allow for complex structures to be easily fabricated using a single mathematical equation. It has high flexibility in describing cellular architectures with the freedom to integrate diverse porous shapes with controlled porosity. One of the most representative ISM methods is the triply periodic minimal surface (TPMS) which has been a topic of great research interest over the past few years. TPMS has been widely used in the replication of natural structures, especially in biological tissues. More details about the TPMS are given in Section 2.

This paper will introduce a new, advanced ISM method using a tetrahedral implicit surface (TIS). A variety of tetrahedral implicit surfaces, including P-, D-, and G-surfaces, are constructed and demonstrated. The major contribution of our work is to address the problem of porous scaffold design from a different perspective based on tetrahedral basic elements, rather than the conventional hexahedrons. The remainder of the paper is organized as follows. In Section 2, we give a brief introduction to TPMSs. Section 3 focuses on the basic mathematical equations of the proposed TIS method and provides a proof to the continuity of TIS surfaces between neighboring tetrahedrons. A flowchart of generating porous scaffolds from a CAD surface model is elaborated on by using the proposed TIS method. In Section 4, various results are compared between different types of TISs and TPMSs. Additionally, some essential properties, such as porosity and mean curvatures, are presented and discussed. Conclusively, a summary of the advantages and deficiencies of TIS is given in Section 5.

2. Triply Periodic Minimal Surfaces

Among various state-of-the-art implicit surface modeling methods, the triply periodic minimal surface has attracted the most attention over the past few decades due to its inherent advantages of geometrical and biological arrangement of pore structure. A minimal surface is locally area-minimizing, which has the smallest area for a surface spanning the boundary of it. The properties of minimal surfaces are very similar to those of a natural biological structure. Many years ago, Galusha proved the existence of minimal surface geometries in intravital biological tissues, including beetle shells and weevils [22,23]. Kapfer also proposed that the TPMS-based sheet solids could provide relatively high stiffness with a high porosity, which creates the possibility of developing structural scaffolds using TPMSs [24]. Later, Rajagopalan first designed coterminous seeding-feeding networks based on Schawartz's Primitive (P) surface to supply nutrients to the proliferating cells [25]. Melchels et al. used CAD softwares to generate TPMS diamond (D) and gyroid (G) architectures, and introduced the gradient in porosity and pore size by adding a linear term for z-values of the Cartesian coordinate system [26]. Based on these sample designs using a specific TPMS in simple cubes, Yoo summarized and presented a comprehensive CAD porous scaffold design methodology using TPMS libraries composed of diverse primitive units [27].

In mathematics, The Enneper-Weierstrass's representation describes a global parametrization of TPMS as in [25]:

$$x = Re(e^{i\theta} \int_{\omega_0}^{\omega} (1 - \tau^2) R(\tau) d\tau)$$

$$y = Re(e^{i\theta} \int_{\omega_0}^{\omega} i(1 + \tau^2) R(\tau) d\tau) \tag{1}$$

$$z = Re(e^{i\theta} \int_{\omega_0}^{\omega} 2\tau R(\tau) d\tau)$$

where $i^2 = -1$, τ is a complex variable and Re means the real part, and θ is the Bonnet angle. More generally, the approximation of TPMS can be expressed as a single-valued function of three variables from the x-, y-, and z-axis. The surface is the locus of points for which the function has a constant value C, which is defined as the threshold of TPMS. The most common TPMS units are listed in Table 1.

Table 1. Mathematical definitions of some common TPMS surfaces.

TPMS Surface Type	Mathematical Definitions
P Surface	$\phi_P(r) = \cos(x) + \cos(y) + \cos(z) = C$
D Surface	$\phi_D(r) = \sin(x)\sin(y)\sin(z) + \sin(x)\cos(y)\cos(z) +$ $\cos(x)\sin(y)\cos(z) + \cos(x)\cos(y)\sin(z) = C$
G Surface	$\phi_G(r) = \sin(x)\cos(y) + \sin(z)\cos(x) + \sin(y)\cos(z) = C$
I-WP Surface	$\phi_{I-WP}(r) = 2[\cos(x)\cos(y) + \cos(y)\cos(z) + \cos(z)\cos(x)]$ $-[\cos(2x) + \cos(2y) + \cos(2z)] = C$

Although TPMS methods can generate a continuous smooth porous network, the continuity and integrality problems may exist when irregular models directly intersect with the iso-surface. As shown in Figure 1, some isolated malformed curves appear on the boundary of the porous femur. Moreover, it is difficult and time-consuming to partition some complex domains into a set of regular or nearly regularly-shaped hexahedrons. To address these problems, it is necessary to introduce a few advanced mapping techniques to interpolate a regular TPMS matrix into a parameterized model mesh. Yoo proposed a heterogeneous porous scaffold design method by combining the distance field with TPMS [28]. Feng et al.

presented a different approach to generating TPMS porous scaffolds based on T-spline [29]. More recently, Chen et al. attempted to map regular TPMS into parameterized hexahedral mesh using the Laplace's equation [30].

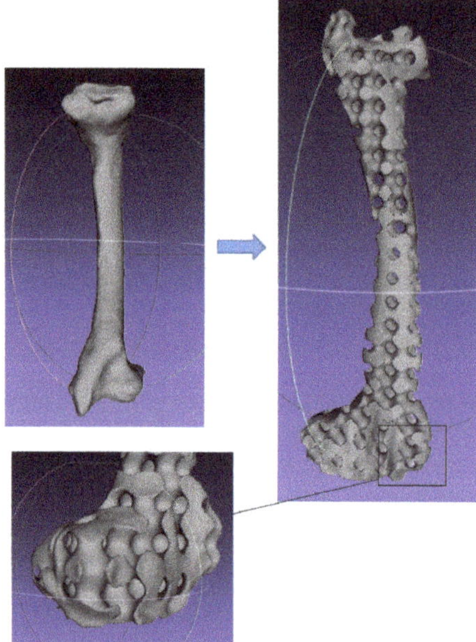

Figure 1. Porous fumer bone scaffold generated by TPMS. Deformed structures are obvious on the surface layer.

Even though the benefits of TPMS and its applications have been widely discussed in the literature, it might not always be the best tool when manufacturing porous structures. The major limitations of hexahedron-based modeling can be summarized from two perspectives:

(1) Quadrilateral and hexahedral meshes are usually less frequently and efficiently used than triangular and tedrahedral meshes in various image-processing areas.
(2) More importantly, a hexahedron will introduce a distortion issue when deformed in 3D which is not seen in a tetrahedron.

3. Tetrahedral Implicit Surface Modeling

3.1. Mathematical Description of Tetrahedral Implicit Surface

TPMS is characterized as a minimal surface which is periodic in three perpendicular directions. Obviously, the same manner does not apply to the generation of periodic surfaces in a tetrahedron. Our goal is to make the surface periodic in certain particular directions, such as the vertical direction of each of the four faces of a tetrahedron. Here, we introduce a new coordinates system using point-to-plane distances. In this system, there are four axes, corresponding to the vertical directions of four faces of the tetrahedron. The coordinate of an arbitrary point inside the tetrahedron on a certain axis is determined by the ratio of the distance from the point to the corresponding face divided by the corresponding height of the tetrahedron. As shown in Figure 2, the coordinates of point P, $(\alpha, \beta, \gamma, \sigma)$, can be calculated by:

$$
\begin{aligned}
P_\alpha &= \frac{d_\alpha}{h_\alpha} \times 2\pi \times \lambda_\alpha \\
P_\beta &= \frac{d_\beta}{h_\beta} \times 2\pi \times \lambda_\beta \\
P_\gamma &= \frac{d_\gamma}{h_\gamma} \times 2\pi \times \lambda_\gamma \\
P_\delta &= \frac{d_\delta}{h_\delta} \times 2\pi \times \lambda_\delta
\end{aligned}
\tag{2}
$$

in which d_α, d_β, d_γ, and d_δ denote the projection distances from point P to plane BCD, ACD, ABD, and ABC, respectively, and h_α, h_β, h_γ, h_δ are the height of the tetrahedron corresponding to each plane. The parameter $\lambda_i (i = \alpha, \beta, \gamma, \delta)$ determines the periodicity of its corresponding direction.

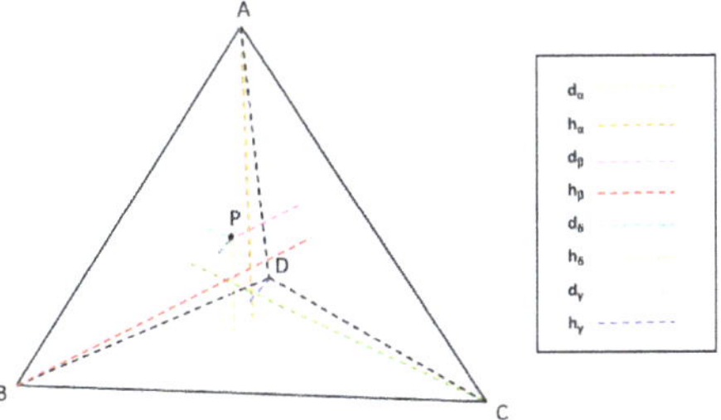

Figure 2. The new coordinates system using scaled point-to-plane distances.

To mimic similar properties of TPMS, we adopt the mathematical expressions of the TPMSs on each periodic component and derive the following TIS equations:

$$
\begin{aligned}
\phi_P(r) &= \cos(\alpha) + \cos(\beta) + \cos(\gamma) + \cos(\delta) \\
&\quad + \cos(\alpha)\cos(\beta)\cos(\gamma)\cos(\delta) = C \\
\phi_D(r) &= \sin(\alpha)\cos(\beta)\cos(\gamma)\cos(\delta) + \cos(\alpha)\sin(\beta)\cos(\gamma)\cos(\delta) \\
&\quad + \cos(\alpha)\cos(\beta)\sin(\gamma)\cos(\delta) + \cos(\alpha)\cos(\beta)\cos(\gamma)\sin(\delta) \\
&\quad + \cos(\alpha)\cos(\beta)\cos(\gamma)\cos(\delta) = C \\
\phi_G(r) &= \cos(\alpha)\sin(\beta) + \cos(\gamma)\sin(\delta) + \cos(\alpha)\sin(\beta) \\
&\quad + \cos(\gamma)\sin(\delta) + \cos(\alpha)\cos(\beta)\cos(\gamma)\cos(\delta) = C
\end{aligned}
\tag{3}
$$

It is worth noting that these TIS surfaces are expressed by implicit functions with a constant threshold C. With a proper threshold given, the TIS surface mesh can be generated as an iso-surface by using the Marching Cubes (MC) algorithm, which is one of the most famous approaches to extracting iso-surfaces from a three-dimension data field.

3.2. Continuity Analysis of Tetrahedral Implicit Surface

As mentioned in Section 2, TPMS couldn't produce a porous structure fitting any irregular complex shape unless some mapping techniques, such as the distance field [31], T-spline [29], or hexahedral parameterization [30], is integrated. Unlike TPMS, a significant advantage of the proposed tetrahedral implicit surface method is that it can generate surfaces with same morphological features inside the tetrahedron of any shape. When multiple tetrahedra are considered, which is almost always true in modeling real-world shapes, the surfaces between neighboring tetrahedrons must be seamlessly connected at the joint. Otherwise, it may cause holes or gaps in the resulting surfaces. In this section, a mathematical geometry proof is provided to validate the C^0 continuity of TIS between two adjacent tetrahedrons.

As shown in Figure 3, $ABCE$ and $ADCE$ are two adjacent tetrahedrons of arbitrary shapes that share the triangular face ACE. F is an arbitrary point on the face ACE. I and J are the perpendicular projections of the point F onto the two faces BCE and DCE, respectively. Similarly, H and K are the projections of the point A onto the two faces BCE and DCE, respectively.

$$because \ FI \perp BCE \ \ and \ \ AH \perp BCE$$
$$therefore, \ FI \parallel AH$$
$$and \ \triangle GFI \sim \triangle GAH \tag{4}$$
$$so \ we \ get: \ r_1 = \frac{GF}{GA} = \frac{GI}{GH} = \frac{FI}{AH}$$

The same holds for tetrahedron $ACDE$ and we have:

$$r_2 = \frac{GF}{GA} = \frac{GJ}{GK} = \frac{FJ}{AK} \tag{5}$$

Combine Equations (4) and (5), then:

$$r = r_1 = r_2 = \frac{GF}{GA} = \frac{FI}{AH} = \frac{FJ}{AK} \tag{6}$$

In tetrahedron $ABCE$, $\frac{FI}{AH} * 2\pi$ and $\frac{FJ}{AK} * 2\pi$ is the coordinate of point F with respect to the faces BCE and CDE, respectively. Similarly, we can conclude that the other two coordinates of point F in tetrahedron $ABCE$ with respect to faces ABC and ABE are the same as those of point F to faces ADC and ADE in tetrahedron $ADCE$. Meanwhile, the coordinate of F to face ACE equals to 0 in both tetrahedrons $ABCE$ and $ADCE$, which means the four coordinates of point F are exactly the same in both of the two tetrahedrons. As a result, we can conclude that the TIS function values of all points lying on the shared face of two neighboring tetrahedrons will be the same, regardless of their shape.

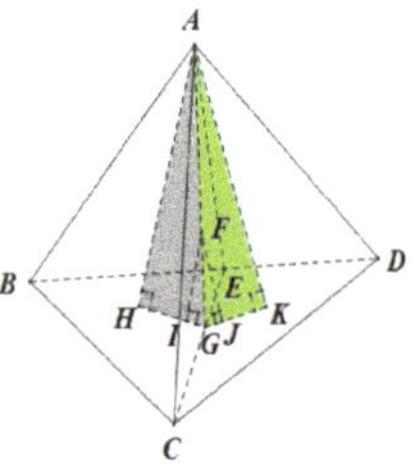

Figure 3. Proof of the continuity of TIS between two adjacent tetrahedrons.

3.3. Porous Structure Generation from Surface Moldes

Based on the proposed TIS method, we can achieve porous scaffolds by tetrahedralizing the tissue with a customized configuration. Biological tissues are inherently heterogeneous architectures. At the macrostructure level, the tissue exhibits great differences in both morphology and biofunctionability. Pore-intensive parts have higher structural strength and can support most of the weight. Conversely, the sparsely populated parts have better permeability and transport of nutrients and metabolic waste. Figure 4 shows a flowchart of the details for designing a 3D heterogeneous porous scaffold from the 3D anatomical shape.

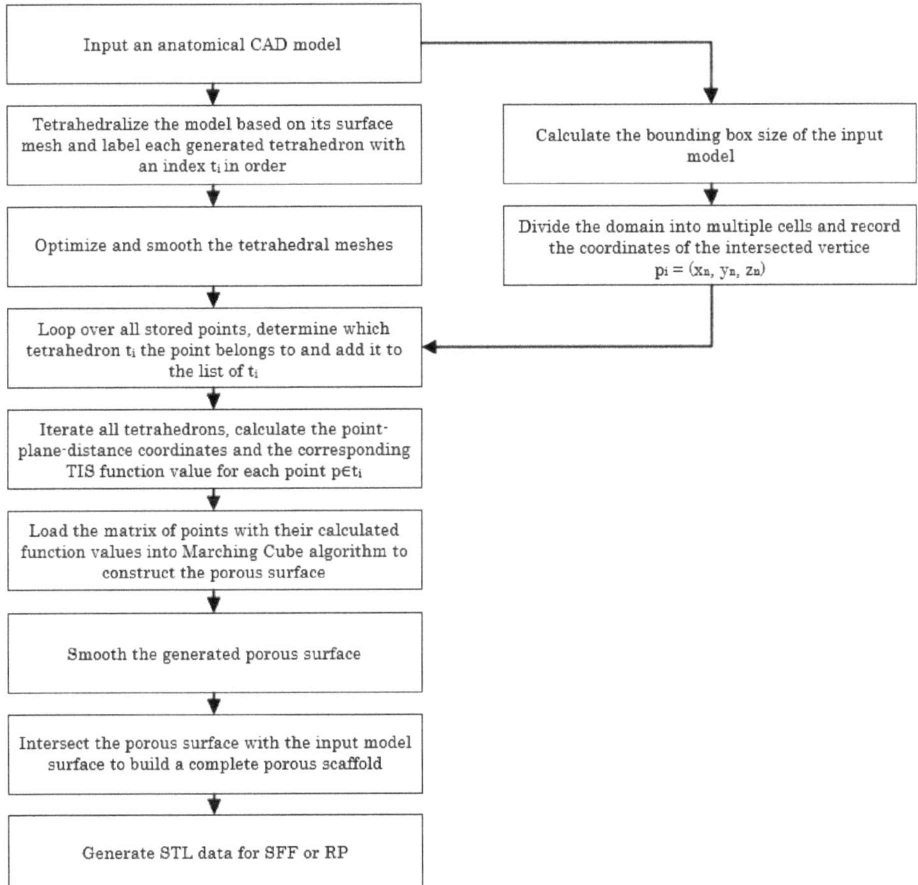

Figure 4. Flowchart showing the procedures for designing a 3D heterogeneous porous scaffold using the TIS method.

Since the quality of the tetrahedral mesh, including the face angle and dihedral angle, has a decisive influence on the quality of the generated scaffold, our approach uses a tetrahedral mesh generation algorithm based on the body-centered cubic (BCC) tetrahedral lattice [32], followed by quality tetrahedral mesh smoothing via boundary-optimized Delaunay triangulation (B-ODT). Compared with the two current popular methods, Tetgen and Netgen, the BCC lattice tetrahedral meshing algorithm is more robust and general and has no quality requirement for the input surface mesh, and the sharp features are also reconstructed better. To tetrahedralize the input model, the octree of the input was firstly subdivided into 3D mesh based on the Euclidean distance transformation. Then, we computed the sign of each node in the BCC lattice and calculated the intersecting points where the edge crossed the input surface mesh. Next, we detected the cutting points that were "too close" to the original BCC nodes and snapped them to the corresponding nodes. Finally, the boundary polyhedra was decomposed into tetrahedrons and the entire generated tetrahedral mesh was post-processed by the B-ODT smoothing algorithm. For more details on the tetrahedral meshing algorithm, please refer to [33,34].

4. Results and Discussion

In this section, various types of TIS surfaces are compared with the TPMS surfaces from different perspectives. The TIS algorithm is implemented in C programming language running on an 8×2.20-GHz Intel(R) Core(TM) i7 computer with 16 GB of memory. A volume of $200 \times 200 \times 200$ voxels was used as the resolution for both TPMS and TIS to generate various types of surfaces. While different selections of the threshold C can significantly change the porosity and pore size, we chose the constant 0 as the threshold of the generation of P-, D-, and G-surfaces for all TPMS models in this paper. For TIS surface generation, however, we used a concept of relative threshold, which is related to the actual threshold as follows:

$$actual\_threshold = (max - min) * relative\_threshold + min \tag{7}$$

in which *max* and *min* are the minimum and maximum of the calculated TIS function values from Equation (3). The range of the relative threshold was rescaled into 0 to 1 according to the actual threshold, which is the real threshold used in the experiments. The periodicity parameter λ, as used in Equation (2), was chosen as follows. To stereoscopically illustrate the fabricated TIS surfaces with a moderate pore size, 0.16 and 0.44 were used as the thresholds for P-surface ($C = 0.16$ when $\lambda = 1$, $C = 0.44$ when $\lambda = 2$ for the generation of TIS surfaces). Also, we used the threshold of 0.67 for the TIS D-surface and a threshold of 0.5 for the TIS G-surface both for $\lambda = 1$ and $\lambda = 2$.

In Figures 5–7, the surfaces in the two left columns are TPMS Schwarz P-, D-, and G- surfaces, while the right two columns contain the TIS surfaces of the same type, respectively. The meshes in the first row were constructed using a single period ($\lambda = 1$), while the meshes in the second row were constructed using a double period ($\lambda = 2$) on each axis instead. From these figures, it is not hard to figure out that the shapes of the surfaces and the locations of the holes are very similar between the same types of surfaces for different groups. For example, the P-surfaces in both TPMS and TIS have an opening on each face of the unit domain (either tetrahedron or hexahedron). The opening on the tetrahedral face is an ellipse of the triangle outline, while the opening on the hexahedral face is a regular circle. The interiors are also connected, which allows for cell growth through the holes on the surface. In addition, Figures 8–10 also give closer views of the surfaces between two neighboring tetrahedrons.

To further analyze the characteristics of these tetrahedral implicit surfaces, we first used the well-known open source CAD software, Meshlab, to compute the average curvatures over the whole surface. Figure 11 demonstrates the distributions of curvatures of different surfaces, as well as their corresponding detailed histograms. The color map ranges from red to blue, indicating the curvature values from a minimum of -30 to a maximum of 30 (the values on the X-axis are curvatures in ascending order, while the values on the Y-axis are the number of points whose curvatures share the same value). Additionally, more information including the minimum/maximum curvature and the mean curvature, as well as the standard deviation of the curvatures are also provided for each of the TPMS and TIS surfaces in Table 2. In the table, the minimum curvatures of the TIS P-surface and G-surface are almost double the values of the TPMS P-surface and G-surface. Only the maximum curvature of the TIS P-surface is significantly larger than the TPMS surface of the same type. Moreover, the standard deviation of the curvatures of TIS surfaces is relatively larger than that of the TPMS surface of the same type. From Figure 11 and Table 2, it is not hard to conclude that the distributions of curvatures of TPMSs are all symmetrical, while this symmetry does not apply to TISs. Nonetheless, one of the most important common attributes for TPMS and TIS is that the overall distribution of the curvatures tends to converge to the mean curvature, which approximates to 0.

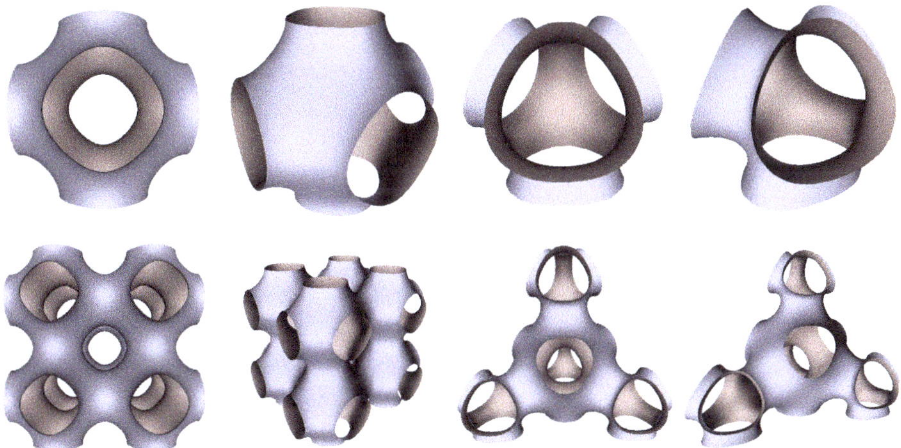

Figure 5. TPMS Schwarz P Surface (left two columns) vs. TIS P Surface (right two columns).

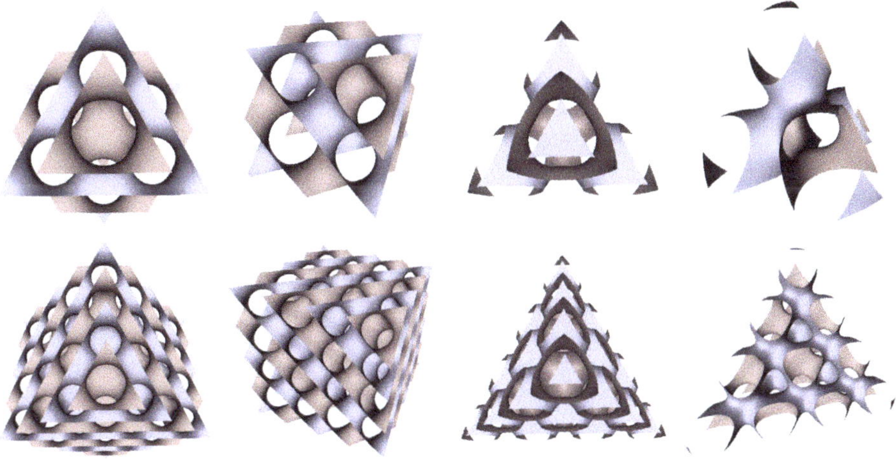

Figure 6. TPMS Schwarz D Surface (left two columns) vs. TIS D Surface (right two columns).

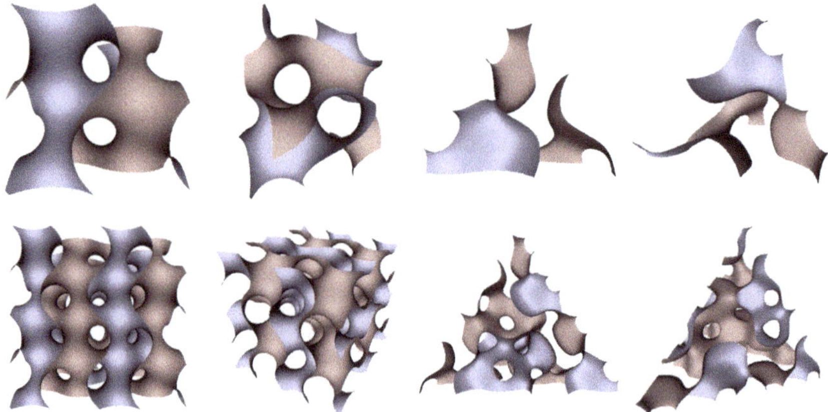

Figure 7. TPMS Schwarz G Surface (left two columns) vs. TIS G Surface (right two columns).

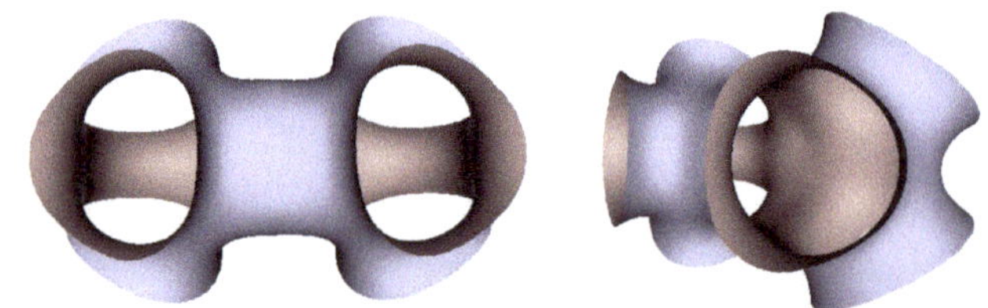

Figure 8. Two views of the TIS P-surface generated from two connecting tetrahedrons.

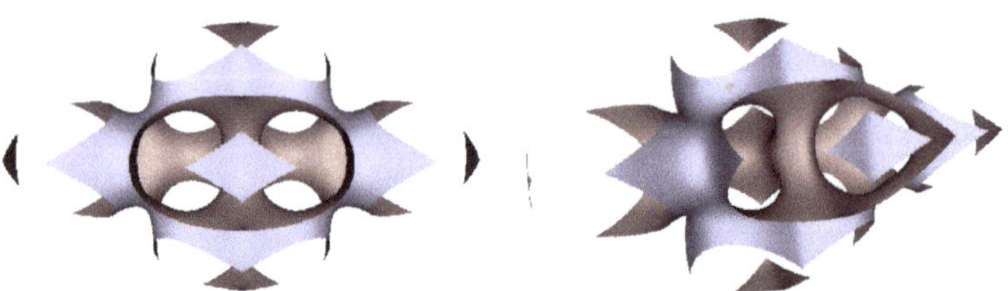

Figure 9. Two views of the TIS D-surface generated from two connecting tetrahedrons.

Besides, we also realized that the porosity and pore size of the surfaces were quite sensitive to the selection of the threshold. Through the experiments, we found that not all thresholds in the range of 0 to 1 could generate a completely connected surface. To create the TIS surfaces of a single period from a regular tetrahedron, the ideal threshold range for a P-surface is 0.118–0.275, for the D-surface is 0.569–0.765, and for the G-surface is 0.431–0.627. A value out of these ranges will cause the generated corresponding type of the deformed surface, which means the hole structure will disappear. On the other hand, TPMS surfaces also have the same issues of surface discontinuity and deformation as TIS surfaces. Nonetheless, the difference is that the selective areas of the reasonable threshold for TPMSs all uniformly concentrate on the range from −1 to 1. Table 3 gives more information of the relationship between measured porosities and different thresholds, C. Additionally, some examples of the deformed TPMSs and TISs with a threshold that exceeds the proper range are demonstrated in Figure 12. It is noted that the hole size gets bigger and the porosity behaves more saturated when C gets increased in the given range for both TPMS and TIS. Meanwhile, the surface areas of different TPMS and TIS surfaces for a unit volume are measured by the sum area of the generated mesh triangles, and the results are presented in Table 4 as a reference. All related parameters, including the resolution and threshold, are the same as mentioned at the beginning of Section 4. In the end, Table 5 lists the running time of the TPMS and TIS algorithms, as well as the number of generated triangles for the P-surfaces with different resolutions. The results indicate that the computation time for the TIS algorithm is relatively shorter and mesh size is much smaller than the TPMS when the same resolution is used.

Figure 10. Two views of the TIS G-surface generated from two connecting tetrahedrons.

Table 2. Statistical analysis to the curvatures of TPMSs and TISs.

Curvatures	TPMS Surfaces			TIS Surfaces		
	P-Surface	D-Surface	G-Surface	P-Surface	D-Surface	G-Surface
Minimum	−12.66	−14.99	−14.34	−33.19	−14.88	−26.48
Maximum	12.59	13.83	18.32	21.09	12.16	15.17
Medium	−0.57	−1.18	−0.66	−1.68	0.45	−1.68
Average	−0.49	−1.29	−0.79	−2.88	−0.94	−2.79
Standard Deviation	3.78	3.32	2.85	4.11	4.32	4.51
Variance	14.28	11.04	8.11	16.88	18.63	20.36

Table 3. Effect of diverse thresholds C on different TPMS and TIS surfaces.

TPMS P-Surface		TPMS D-Surface		TPMS G-Surface	
Threshold C	Porosity (%)	Threshold C	Porosity (%)	Threshold C	Porosity (%)
−0.95	22.79	−0.95	10.28	−0.95	18.77
−0.5	35.71	−0.5	29.42	−0.5	33.81
0	49.98	0	49.94	0	50.00
0.5	64.28	0.5	70.57	0.5	66.19
0.95	77.21	0.95	89.71	0.95	81.22
TIS P-Surface		**TIS D-Surface**		**TIS G-Surface**	
Threshold C	Porosity (%)	Threshold C	Porosity (%)	Threshold C	Porosity (%)
0.118	18.36	0.569	57.35	0.431	44.03
0.157	35.80	0.618	66.88	0.480	57.29
0.196	49.36	0.667	75.73	0.529	67.76
0.235	61.17	0.716	83.04	0.578	76.09
0.275	71.86	0.765	89.90	0.627	84.04

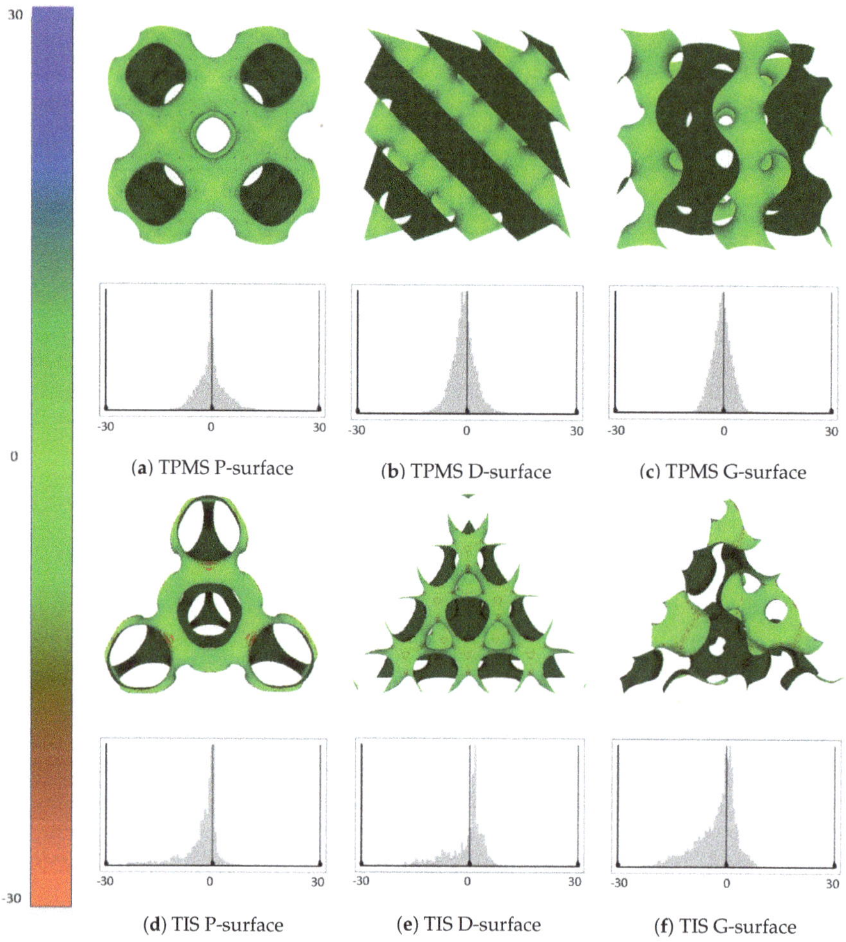

(**a**) TPMS P-surface (**b**) TPMS D-surface (**c**) TPMS G-surface

(**d**) TIS P-surface (**e**) TIS D-surface (**f**) TIS G-surface

Figure 11. Distribution maps of the mean curvature and corresponding histograms for TPMSs and TISs.

Table 4. Surface areas of TPMSs and TISs for a unit volume.

Surface Type		Surface Area for Unit Volume
TPMS Surfaces	P-Surface	2.353180
	D-Surface	3.828864
	G-Surface	3.083422
TIS Surfaces	P-Surface	3.798055
	D-Surface	2.367160
	G-Surface	2.654556

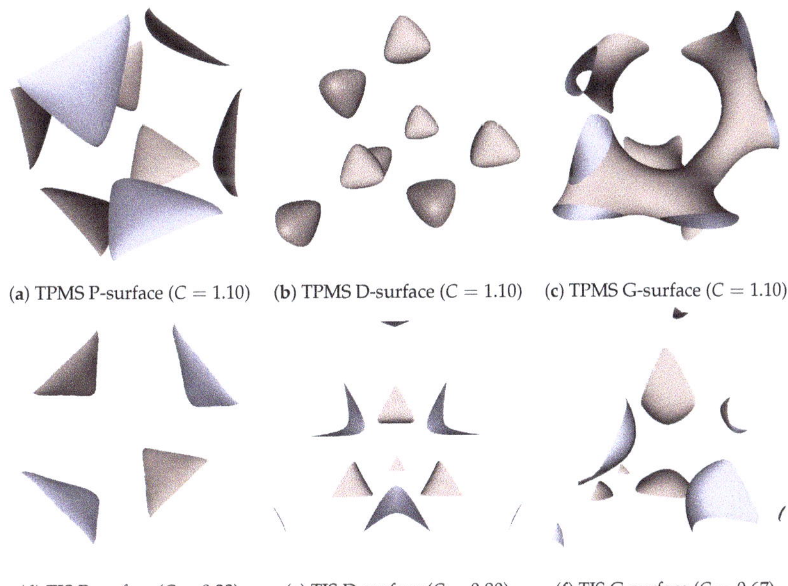

(**a**) TPMS P-surface ($C = 1.10$)   (**b**) TPMS D-surface ($C = 1.10$)   (**c**) TPMS G-surface ($C = 1.10$)

(**d**) TIS P-surface ($C = 0.33$)   (**e**) TIS D-surface ($C = 0.80$)   (**f**) TIS G-surface ($C = 0.67$)

Figure 12. Deformed TPMS and TIS surfaces with a threshold C that exceeds the reasonable range.

Table 5. Running time comparison between TPMS and TIS.

Resolution	TPMS P-Surface		TIS P-Surface	
	Computational Time (s)	Number of Triangles	Computational Time (s)	Number of Triangles
100	0.108	74,216	0.083	27,379
200	1.212	299,672	0.674	112,381
300	3.895	675,848	2.930	254,074
400	9.749	1,203,704	7.361	454,630
500	18.524	1,882,376	14.708	708,986

Numerous experimental results implied that the TIS P-surface performs better in the construction of large models than the D- and G-surfaces. Below, we use the P-surface as an example to elaborate the processes of the generation of a porous femur scaffold. The input model is a 3D femur mesh composed of 7587 vertices and 15,174 faces as shown in Figure 13. Through the tetrahedral meshing operation, the femur model is firstly tiled into 989 tetrahedrons, in which over 95 percents of triangular angles lies in the range between 50 degrees and 100 degrees. Figure 14a shows the surface mesh of the tetrahedralized femur model and Figure 14b presents the cross-sectional view of the generated tetrahedral mesh. Secondly, by calculating the TIS function values for all grid points inside the tetrahedral network, we can easily extract the iso-surface through the Marching Cubes algorithm using the threshold $C = 0.176$. Moreover, we applied Laplacian smoothing and curvature-based smoothing on the entire porous surface mesh to improve the C^1 continuity at the joints. Figure 15 depicts the smoothed porous TIS surface constrained by the input femur's surface mesh. Lastly, we intersected the original input femur model with the constructed porous TIS volume to produce a porous femur scaffold, as shown in Figure 16.

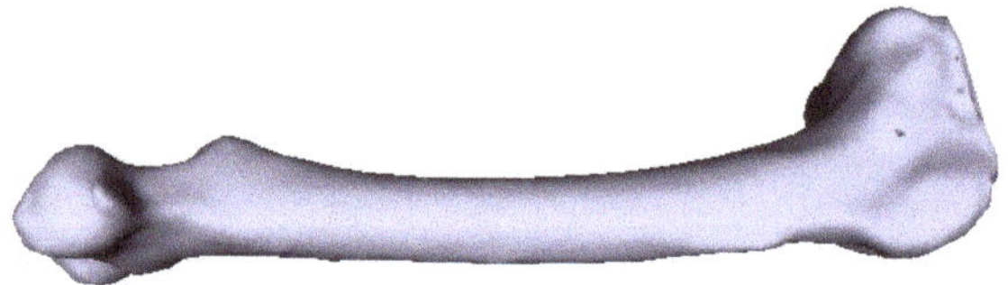

Figure 13. Original 3D femur model as input.

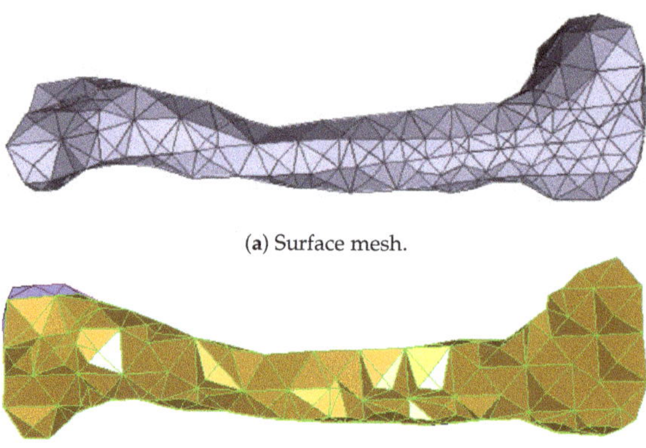

(**a**) Surface mesh.

(**b**) Cross-section view.

Figure 14. Tetrahedral meshes obtained from the input femur model.

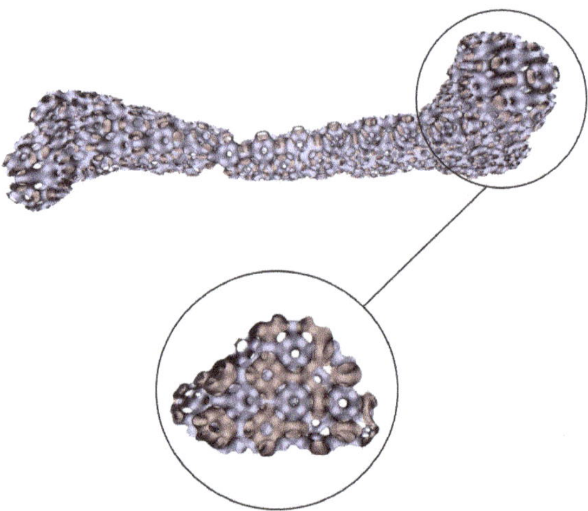

Figure 15. Porous TIS surface constrained by the femur's profile.

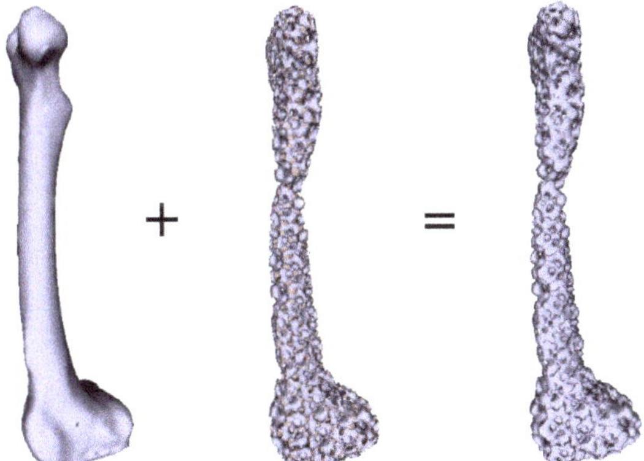

Figure 16. Intersection of the original femur volume with the generated porous volume producing a porous femur scaffold.

5. Conclusions and Future Work

In this paper, an implicit 3D surface method for designing bio-scaffolds based on the tetrahedral implicit surface was proposed. To the best of our knowledge, this is the first attempt to construct porous structures based on tetrahedral elements. The TIS method demonstrated its great potential in the field of designing optimized porous structures. Compared to the well-established TPMS approach, this concise method brings several advantages to the tissue engineering scaffold design:

(1) Generation of the TIS surface could avoid the restriction of modeling based on a regular unit domain, such as the regular hexahedron in TPMS. Even in the deformed tetrahedron, there can still be the creation of a characteristic tetrahedral surface, which can guarantee C^0 continuity between adjacent TIS surfaces.
(2) Different from the TPMS method that needs to parametrize tissue surfaces into hexahedral meshes, it is more convenient to tetrahedralize the triangular surface mesh as in the proposed TIS method. Without the special mapping procedure, the entire process of generating pores is also simplified.
(3) The strong interconnectivity and tetrahedron-based modeling grants the TIS more flexibility and creativity for complicated shape modeling.

Notwithstanding this, there are also several limitations of this method which needs further research. Though the TIS surface of multiple periods obtained from a single unit has shown the characteristic of approximately zero mean curvature, and though we also showed the C^0 continuity of the generated surfaces, the C^1 continuity of the surface at the junction of two neighboring tetrahedrons is not guaranteed and needs further investigation. Currently, the mesh smoothing algorithm is used to greatly improve the smoothness of the resulting surface. Moreover, some isolated closed surfaces may be produced in the TIS D-surface from a complex input model.

Author Contributions: Conceptualization, Y.G. and Z.Y.; methodology, Y.G. and Z.Y.; software, Y.G.; validation, Y.G., K.L. and Z.Y.; formal analysis, Y.G.; investigation, K.L.; resources, Z.Y.; data curation, Z.Y.; writing—original draft preparation, Y.G.; writing—review and editing, Y.G. and Z.Y.; visualization, Y.G.; supervision, Z.Y.

Funding: This research received no external funding.

Conflicts of Interest: The authors declare no conflict of interest.

References

1. Mikos, A.G.; Sarakinos, G.; Leite, S.M.; Vacanti, J.P.; Langer, R. Laminated three dimensional biodegradable foams for use in tissue engineering. *Biomaterials* **1993**, *14*, 323–330, doi:10.1016/B978-008045154-1.50013-7. [CrossRef]

2. Mooney, D.J.; Baldwin, D.F.; Suh, N.P.; Vacanti, J.P.; Langer, R. Novel approach to fabricate porous sponges of poly(D,L-lactic-*co*-glycolic acid) without the use of organic solvents. *Biomaterials* **1996**, *17*, 1417–1422, doi:10.1016/0142-9612(96)87284-X. [CrossRef]

3. Guan, J.; Fujimoto, K.L.; Sacks, M.S.; Wagner, W.R. Preparation and characterization of highly porous, biodegradable polyurethane scaffolds for soft tissue applications. *Biomaterials* **2005**, *26*, 3961–3971, doi:10.1016/j.biomaterials.2004.10.018. [CrossRef]

4. Van de Witte, P.; Dijkstra, P.J.; van den Berg, J.W.A.; Feijen, J. Phase behavior of polylactides in solvent–nonsolvent mixtures. *J. Polym. Sci.* **1996**, *34*, 2553–2568, doi:10.1002/(SICI)1099-0488(19961115)34:15<2553::AID-POLB3>3.0.CO;2-U. [CrossRef]

5. Yoshikawa, H.; Tamai, N.; Murase, T.; Myoui, A. Interconnected porous hydroxyapatite ceramics for bone tissue engineering. *J. R. Soc. Interface* **2009**, *6*, S341–S348, doi:10.1098/rsif.2008.0425.focus. [CrossRef] [PubMed]

6. Sill, T.J.; von Recum, H.A. Electrospinning: Applications in drug delivery and tissue engineering. *J. R. Soc. Interface* **2008**, *29*, 1989–2006, doi:10.1016/j.biomaterials.2008.01.011. [CrossRef] [PubMed]

7. Martins, A.; Duarte, A.R.C.; Faria, S.; Marques, A.P.; Reis, R.L.; Neves, N.M. Osteogenic induction of hBMSCs by electrospun scaffolds with dexamethasone release functionality. *J. R. Soc. Interface* **2010**, *31*, 5875–5885, doi:10.1016/j.biomaterials.2010.04.010. [CrossRef]

8. Knackstedt, M.A.; Arns, C.H.; Senden, T.J.; Gross, K. Structure and properties of clinical coralline implants measured via 3D imaging and analysis. *Biomaterials* **2006**, *27*, 2776–2786, doi:10.1016/j.biomaterials.2005.12.016. [CrossRef] [PubMed]

9. Tampieri, A.; Sprio, S.; Ruffini, A.; Celotti, G.; Lesci, I.G.; Roveri, N. From wood to bone: multi-step process to convert wood hierarchical structures into biomimetic hydroxyapatite scaffolds for bone tissue engineering. *J. Mater. Chem.* **2009**, *19*, 4973–4980, doi:10.1039/B900333A. [CrossRef]

10. Lo, H.; Ponticiello, M.S.; Leong, K.W. Fabrication of controlled release biodegradable foams by phase separation. *Tissue Eng.* **1995**, *1*, 15–28, doi:10.1089/ten.1995.1.15. [CrossRef]

11. Melchels, F.P.W.; Domingos, M.A.N.; Klein, T.J.; Malda, J.; Bartolo, P.J.; Hutmacher, D.W. Additive manufacturing of tissues and organs. *Prog. Polym. Sci.* **2012**, *37*, 1079–1104, doi:10.1016/j.progpolymsci.2011.11.007. [CrossRef]

12. Bartolo, P.J.S.; Almeida, H.; Laoui, T. Rapid prototyping and manufacturing for tissue engineering scaffolds. *Int. J. Comput. Appl. Technol.* **2009**, *36*, 1–9, doi:10.1504/IJCAT.2009.026664. [CrossRef]

13. Peltola, S.M.; Melchels, F.P.; Kellomäki, M. A review of rapid prototyping techniques for tissue engineering purposes. *Ann. Med.* **2008**, *40*, 268–280, doi:10.1080/07853890701881788. [CrossRef] [PubMed]

14. Giannitelli, S.; Accoto, D.; Trombetta, M.; Rainer, A. Current trends in the design of scaffolds for computer-aided tissue engineering. *Acta Biomater.* **2014**, *10*, 580–594, doi:10.1016/j.actbio.2013.10.024. [CrossRef] [PubMed]

15. Lee, K.W.; Wang, S.; Dadsetan, M.; Yaszemski, M.J.; Lu, L. Enhanced Cell Ingrowth and Proliferation through Three-Dimensional Nanocomposite Scaffolds with Controlled Pore Structures. *Biofabrication* **2010**, *11*, 682–689, doi:10.1021/bm901260y. [CrossRef] [PubMed]

16. Duan, B.; Cheung, W.L.; Wang, M. Optimized fabrication of Ca–P/PHBV nanocomposite scaffolds via selective laser sintering for bone tissue engineering. *Biofabrication* **2011**, *3*, 015001. [CrossRef] [PubMed]

17. Chiu, W.K.; Yeung, Y.C.; Yu, K.M. Toolpath generation for layer manufacturing of fractal objects. *Rapid Prototyp. J.* **2006**, *12*, 214–221, doi:10.1108/13552540610682723. [CrossRef]

18. Sun, W.; Starly, B.; Nam, J.; Darling, A. Bio-CAD modeling and its applications in computer-aided tissue engineering. *Comput.-Aided Des.* **2005**, *37*, 1097–1114, doi:10.1016/j.cad.2005.02.002. [CrossRef]

19. Bucklen, B.S.; Wettergreen, W.A.; Yuksel, E.; Liebschner, M.A.K. Bone-derived CAD library for assembly of scaffolds in computer-aided tissue engineering. *Virtual Phys. Prototyp.* **2008**, *3*, 13–23, doi:10.1080/17452750801911352. [CrossRef]

20. Hollister, S.J.; Levy, R.A.; Chu, T.M.; Halloran, J.W.; Feinberg, S.E. An Image-Based Approach for Designing and Manufacturing Craniofacial Scaffolds. *Int. J. Oral Maxillofac. Surg.* **2000**, *29*, 67–71. [CrossRef]
21. Smith, M.H.; Flanagan, C.L.; Kemppainen, J.M.; Sack, J.A.; Chung, H.; Das, S.; Hollister, S.J.; Feinberg, S.E. Computed Tomography-Based Tissue-Engineered scaffolds in craniomaxillofacial surgery. *Int. J. Med. Robot. Comput. Assist. Surg. MRCAS* **2007**, *3*, 207–216. [CrossRef] [PubMed]
22. Galusha, J.W.; Richey, L.R.; Gardner, J.S.; Cha, J.N.; Bartl, M.H. Discovery of a diamond-based photonic crystal structure in beetle scales. *Phys. Rev. E* **2008**, *77*, 050904, doi:10.1103/PhysRevE.77.050904. [CrossRef] [PubMed]
23. Galusha, J.W.; Richey, L.R.; Jorgensen, M.R.; Gardner, J.S.; Bartl, M.H. Study of natural photonic crystals in beetle scales and their conversion into inorganic structures via a sol-gel bio-templating route. *J. Mater. Chem.* **2010**, *20*, 1277–1284, doi:10.1039/B913217A. [CrossRef]
24. Kapfer, S.C.; Hyde, S.T.; Mecke, K.; Arns, C.H.; Schröder-Turk, G.E. Minimal surface scaffold designs for tissue engineering. *Biomaterials* **2011**, *32*, 6875–6882, doi:10.1016/j.biomaterials.2011.06.012. [CrossRef]
25. Rajagopalan, S.; Robb, R.A. Schwarz meets Schwann: Design and fabrication of biomorphic and durataxic tissue engineering scaffolds. *Med. Image Anal.* **2006**, *10*, 693–712, doi:10.1016/j.media.2006.06.001. [CrossRef]
26. Melchels, F.P.; Bertoldi, K.; Gabbrielli, R.; Velders, A.H.; Feijen, J.; Grijpma, D.W. Mathematically defined tissue engineering scaffold architectures prepared by stereolithography. *Biomaterials* **2010**, *31*, 6909–6916, doi:10.1016/j.biomaterials.2010.05.068. [CrossRef]
27. Yoo, D.J. Computer-aided porous scaffold design for tissue engineering using triply periodic minimal surfaces. *Int. J. Precis. Eng. Manuf.* **2011**, *12*, 61–71, doi:10.1007/s12541-011-0008-9. [CrossRef]
28. Yoo, D.J. Porous scaffold design using the distance field and triply periodic minimal surface models. *Biomaterials* **2011**, *32*, 7741–7754, doi:10.1016/j.biomaterials.2011.07.019. [CrossRef]
29. Feng, J.; Fu, J.; Shang, C.; Lin, Z.; Li, B. Porous scaffold design by solid T-splines and triply periodic minimal surfaces. *Comput. Methods Appl. Mech. Eng.* **2018**, *336*, 333–352, doi:10.1016/j.cma.2018.03.007. [CrossRef]
30. Chen, H.; Guo, Y.; Rostami, R.; Fan, S.; Tang, K.; Yu, Z. Porous Structure Design Using Parameterized Hexahedral Meshes and Triply Periodic Minimal Surfaces. In Proceedings of the Computer Graphics International 2018, Bintan Island, Indonesia, 11–14 June 2018; pp. 117–128, doi:10.1145/3208159.3208174. [CrossRef]
31. Yang, N.; Quan, Z.; Zhang, D.; Tian, Y. Multi-morphology transition hybridization CAD design of minimal surface porous structures for use in tissue engineering. *Comput.-Aided Des.* **2014**, *56*, 11–12, doi:10.1016/j.cad.2014.06.006. [CrossRef]
32. Molino, N.; Bridson, R.; Teran, J.; Fedkiw, R. A crystalline, red green strategy for meshing highly deformable objects with tetrahedra. In Proceedings of the 12th International Meshing Roundtable, Santa Fe, NM, USA, 14–17 September 2003; pp. 103–114.
33. Wang, J.; Yu, Z. Feature-sensitive tetrahedral mesh generation with guaranteed quality. *Comput.-Aided Des.* **2012**, *44*, 400–412, doi:10.1016/j.cad.2012.01.002. [CrossRef] [PubMed]
34. Gao, Z.; Yu, Z.; Holst, M. Quality Tetrahedral Mesh Smoothing via Boundary-Optimized Delaunay Triangulation. *Comput. Aided Geom. Des.* **2012**, *29*, 707–721, doi:10.1016/j.cagd.2012.07.001. [CrossRef] [PubMed]

Article

Topology Optimization for Additive Manufacturing as an Enabler for Light Weight Flight Hardware

Melissa Orme [1,*], Ivan Madera [1], Michael Gschweitl [2] and Michael Ferrari [2]

[1]   Morf3D Inc., El Segundo, CA 90245, USA; ivan@morf3d.com
[2]   Ruag Space, 8052 Zürich, Switzerland; michael.gschweitl@ruag.com (M.G.); michael.ferrari@ruag.com (M.F.)
*   Correspondence: melissa@morf3d.com; Tel.: +1-310-607-0188

Received: 24 September 2018; Accepted: 19 November 2018; Published: 25 November 2018

Abstract: Three case studies utilizing topology optimization and Additive Manufacturing for the development of space flight hardware are described. The Additive Manufacturing (AM) modality that was used in this work is powder bed laser based fusion. The case studies correspond to the redesign and manufacture of two heritage parts for a Surrey Satellite Technology LTD (SSTL) Technology Demonstrator Space Mission that are currently functioning in orbit (case studies 1 and 2), and a system of five components for the SpaceIL's lunar launch vehicle planned for launch in the near future (case study 3). In each case, the nominal or heritage part has undergone topology optimization, incorporating the AM manufacturing constraints that include: minimization of support structures, ability to remove unsintered powder, and minimization of heat transfer jumps that will cause artifact warpage. To this end the topology optimization exercise must be coupled to the Additive Manufacturing build direction, and steps are incorporated to integrate the AM constraints. After design verification by successfully passing a Finite Element Analysis routine, the components have been fabricated and the AM artifacts and in-process testing coupons have undergone verification and qualification testing in order to deliver structural components that are suitable for their respective missions.

Keywords: additive manufacturing; topology optimization; design for additive manufacturing; 3D printing; aerospace; full-life cycle manufacturing flow

---

1. Introduction

Additive Manufacturing, or AM, is the potentially disruptive manufacturing technology in which a structural component is fabricated layer by layer via digital information. It is termed 'Additive' because material is added to every cross sectional layer, as opposed to 'Subtractive' in which material is removed from a raw stock. The aerospace industry is an industrial sector that has much to gain from Additive Manufacturing [1]. Fabricating structural components layer by layer from digital information provides the benefits of: increased design freedom, including the ability to exploit the results of topology optimization algorithms to significantly lightweight parts; eliminating system part count by consolidating assemblies into fewer parts; reducing the lifecycle time between concept, design, manufacture, and validated component delivery; and, a buy-to-fly ratio that approaches unity, thereby leading to significant raw material cost savings as well as leaving a positive environmental impact. Moreover, AM also has the ability to incorporate added functionality, such as internal cooling channels and thermowells into components; manufacture highly complex geometries that would be impossible or very difficult to be realized with traditional methods; and, to combat component obsolescence by fabricating on demand [2].

This paper discusses the design for Additive Manufacturing of light-weight, topologically optimized components that are intended for spaceflight, and will focus on three illustrative examples:

the first two of which are currently found on an orbiting SSTL satellite [3], and the third that is an engine support structure for a lunar launch vehicle [4]. Because the components must be qualified for flight, the designs are verified by rigorous Finite Element Modeling analyses. Subsequently, the Additively Manufactured artifacts and in-process witness coupons undergo materials and mechanical testing in order to verify the required microstructural characteristics and mechanical properties. Because Additive Manufacturing considerations, such as the minimization of overhangs, encapsulated powder, and heat accumulation are not included in the topology optimization formulations that are presented in this work, a subsequent Finite Element Modeling is necessary because the aforementioned considerations are incorporated manually in the topology optimization routine with significant 'human-in-the-loop' iterative analysis. The manual step also includes incorporating allowances for tool access, assembly, fixturing, and application of testing sensors, which are refinements that could occur for any mode of manufacturing (e.g., traditional or AM).

2. Background

The past decade has seen the technology of Solid Freeform Fabrication, a term that was coined in 1987 to describe the fabrication of structural freeform components directly from a computer model without part-specific tooling [5], gain considerable traction in industry. Currently, the term 'Additive Manufacturing' has evolved to generally describe the technologies, such as Solid Freeform Fabrication, which strives to build functional components layer-by-layer from digital information, and includes powder bed fusion processes, wire deposition, and directed energy deposition [6]. Its traditional counterpart, 'Subtractive Manufacturing', encompasses conventional manufacturing methods in which material is successively removed from a bulk with methods such as CNC machining in order fabricate a structural component.

One of the more successful technologies of AM is that of Selective Laser Melting, which is a powder bed fusion process [7,8], as illustrated schematically in Figure 1. The main elements of the Selective Laser Melting process are a leveled bed of metal powder and a focused laser beam that scans the surface of the powder bed according to a digital representation of the desired component, causing localized melting and subsequent rapid solidification. The CAD file of the desired part is converted into an STL file and is subsequently 'sliced' into thin cross-sectional layers. Once the laser scans a layer of powder according to the sliced STL file, the powder bed is physically lowered (indicated by the downward orange arrow) a selectable distance in the range of 30–90 microns [1.2–3.5 thousandths of an inch], and a new layer of powder is swept onto the top of the powder bed from the powder dispenser by the recoater blade. The laser beam then scans the new layer with information corresponding to the new geometrical slice and the process is repeated until the three-dimensional (3D) component is fabricated and contained within the lowered powder bed. To retrieve the fabricated component, the powder bed is then raised and excess powder is removed and sieved for recycling if the specific Additive program allows for using recycled powder. It has been shown experimentally that there are no significant differences in elemental concentrations between virgin powder and that recycled powder after eight builds [9]. However, in many cases, flight-worthy hardware is required to be fabricated from virgin powder, leaving the recycled powder to be used for prototypes. More detailed descriptions of Additive Manufacturing can be found elsewhere [10–15]. An important consequence of Selective Laser Melting is that due to the 'selective', or localized aspect of the melting and subsequent rapid solidification, the resulting microstructure is significantly refined [16], leading to enhanced mechanical properties, such as ultimate strength and yield strength [17].

The enhanced mechanical properties inherent to rapid solidification that are associated with laser based fusion are reduced, however, if the fabricated part is subjected to subsequent heat treatment, which is often employed in an effort to remove internal residual stresses that accumulate due to the anisotropy of the heat transfer associated with Additively Manufactured components [18–20]. For this reason, it is important to consider heat transfer issues in the design phase, which is intrinsically coupled with the orientation of the layerwise build in an effort to reduce, when possible, anisotropy of heat transfer that is associated with the building phase of the part fabrication.

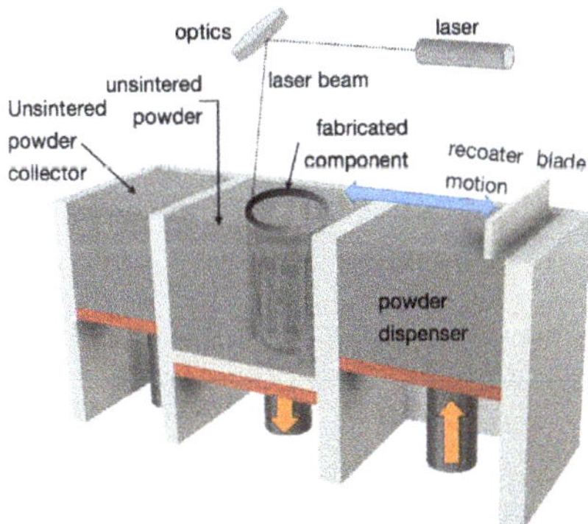

Figure 1. Conceptual schematic of powder bed laser Additive Manufacturing.

Additionally, it is often wrongly assumed that Hot Isostatic Pressing (HIP) need be performed on aluminum alloys in an effort to reduce porosity within powder bed fusion components. Performing HIP on the components will result in even further reductions in strength than stress relieving treatments. It should be noted that all AlSi10Mg components that are reported in this work have an as-built (un heat-treated or post processed) density of 99.7% or higher, which corroborates other literature values of of 99.8% [21], making the need for HIP for the purpose of reducing porosity in AlSi10Mg components moot.

3. Holistic Process Flow

In order to Additively Manufacture components that are worthy of spacecraft or aircraft flight, it is necessary that they be created with reliable and repeatable mechanical and material properties. To this end, a holistic process flow has been developed and established to ensure reliable and repeatable AM parts. The holistic process flow, which is illustrated in Figure 2 includes: candidate part selection; topology optimization for Additive Manufacturing, including AM design considerations (e.g., support structure removal, powder removal, minimization of residual stresses through support structure design, and part design); finite element modeling for analytic verification; AM fabrication of the artifacts and in-process testing coupons; and, final verification at both coupon and component levels. With two separate feedback loops, the holistic process-flow includes verification steps that are incorporated to insure that: (a) the optimized design can withstand the specified loading conditions (first feedback loop) and (b) the mechanical integrity and performance of the manufactured component is suitable for conditions in which it will be employed (second feedback loop). The process flow is discussed in more detail elsewhere [3,4,22]. Note that the topology optimization routine that is employed in this work is the commercially available Altair Hyperworks software package, which, at the time of this work, did not include the AM considerations described above, and hence these are included in manual iterative loops. The holistic process flow refers not to the Topology Optimization exercise, but rather to the entire flow from candidate part selection to verification for flight.

3.1. Candidate Part Selection

Because a component can be additively manufactured does not mean that it is necessarily suitable. To benefit from Additive Manufacturing, added value should be found in terms of reducing weight, manufacturing lead time, part consolidation, added functionality, added complexity, or combating obsolescence. Not all flight components are suitable for Additive Manufacturing. The case studies that are discussed in this chapter have been identified as suitable candidates for Additive Manufacturing due to the fact that they will be significantly lighter in weight than their subtractive counterparts, and will be fabricated on a rapid timescale (concept to validation in eight weeks). The test cases presented have been purposely selected because they have the potential to experience significant weight reductions over their nominal counterparts. Not all AM topologically optimized designs will experience the same weight reduction.

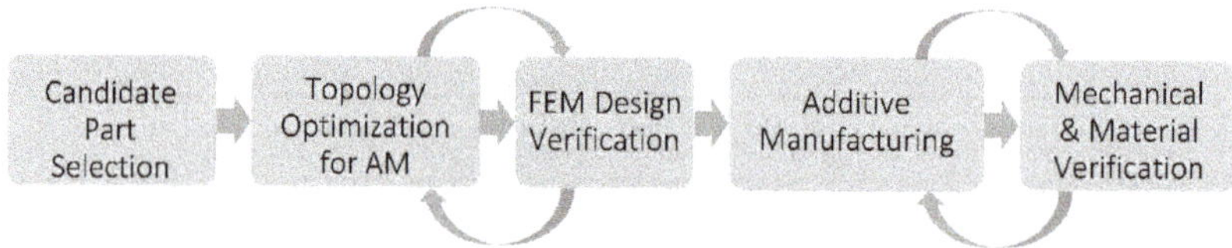

Figure 2. Holistic Process Flow for the fabrication of repeatable AM components.

3.2. Topology Optimization for Additive Manufacturing

The discipline of topology optimization has been used in industry for over two decades [23] and it describes a mathematical method to determine the optimum distribution of material in a design space for a given set of loading and boundary conditions. Early application of topology optimization was found on General Motors Powertrain for under hood brackets [24] and was utilized with subtractive manufacturing.

Until recently, manufacturing constraints limited the utilization of topology optimization because conventional manufacturing could not always produce a topologically optimized component. The advent of AM, however, has revealed renewed interest in the discipline of topology optimization, because many topologically optimized components can now be made with AM that were not possible otherwise [25,26].

Because the topology optimization exercise removes material from all locations where it is not necessary to support the specific loads or satisfy specific boundary conditions, resulting components often contain structures that are not constant in cross section and resemble tree branches or bones, and hence, are termed 'bionic' or 'organic'. The fabrication of hollow structures, structures with internal cooling channels, organic, bionic shaped structures, and structures filled with lattice elements [27] can now be made via Additive Manufacturing.

An important aspect of the topologically optimized design for Additive Manufacturing is to create self-supporting components, or when not possible, components with the minimal number of support structures. Support structures are those sacrificial elements that are incorporated into the component build that support low angled members with respect to the build plane. A good rule of thumb is that members or components greater than 45 degrees from the build plate platform can be printed without support structures. Angles as low as 27 degrees have also been achieved depending on the aspect ratio of the member in question. Hence the designs are driven in part by component orientation with respect to build direction to ensure minimal support structures. Additionally, support structures may be added to mitigate thermal effects that can be detrimental for certain geometries. Hence, design efforts focus on creating approximately uniform member cross sections in an effort to achieve nearly homogeneous heat transfer towards the base plate, thereby minimizing the occurrence of thermal related warping. Furthermore, since downward facing surfaces are known to exhibit inferior characteristics in terms of surface quality and hence dynamic performance, these regions not chosen for highly loaded areas. Downward facing surfaces, termed 'downskins', are typically more rough than the upward facing surfaces due to the fact that those surfaces are in contact with unmelted powder, whereas upskins are in contact with a solid. Because heat conduction through a powder is different than that of a solid, the phase change characteristics of those two regions are markedly different, resulting in different surface textures. The relatively rough downskins can be improved, however, by altering the processing parameters (e.g., laser power, speed, and hatch separation) in the downskin regions to account for the less efficient heat conduction through the powder.

In this work, all topology optimization efforts were conducted in collaboration with RUAG Space and were employed with Altair's Hyperworks 14.0 software suite in order to determine the optimum placement of material with respect to the specific loading and geometric requirements. In order to set-up the topology optimization exercise, interface locations (locations where the part is mounted to the spacecraft or another unit is mounted to the part), as well as stay-in volumes and non-design zones are specified to obtain a resulting design volume. The stay-in volume is the envelope in which the part needs to remain in order to avoid interference with neighboring components on the spacecraft or to allow suitable clearance for part integration or tool access. Non-design zones correspond to fixed regions where material must remain and may correspond to attachment points, or other inherent features of the component. The resulting design space (stay-in volume minus non-design space) is a volume in which the topology optimization algorithm decides where material is needed to fulfill the structural requirements of the part.

Additive Manufacturing considerations that arise due to the layerwise fabrication, such as geometric distortion due to thermal jumps, anisotropic heat accumulation as a function of component geometry coupled with laser scanning regime, the need to support overhanging members, and the need to remove unsintered powder are not intrinsically included in the topology optimization formulations presented herein, however they are considered through manual 'human-in-the-loop' iterations.

Several exciting works have focused on including the AM constraints, such as support structure elimination (through overhang elimination) in topology optimization routines and are included in references [28–36]. Additionally, Liu et al. [37] recently presented a review paper on recent trends and future challenges that are associated with the incorporation of AM constraints and considerations into topology optimization algorithms. The aforementioned references are computational in nature, focused on the development of valid topology optimization routines for AM, and do not focus on the full value-stream of additive manufacturing.

3.3. FEM Design Verification

All analysis and optimization in this work was performed with Altair's Hyperworks 14.0, which uses Hypermesh as pre-processor, OptiStruct as solver, and Hyperview for post-processing. A convergence study was not carried out in this work; however, the results are validated through a material test campaign that is the final step of the holistic process flow and is described in subsequent sections of this work. The material is Additively Manufactured aluminum alloy AlSi10Mg, which is assumed to behave as a linear elastic material. The material properties from previous AM material characterization studies have been employed in this effort.

The mesh size in the analysis has been constrained to be relatively small in order to properly assess stress concentrations (approximately 1 million elements and around 300,000 Nodes, with a mesh size of approx. 1.2–2.5 mm [47–98 thousandths of an inch]). The AM parts are modeled with solid elements (tetra).

Due to the fact that Additive Manufacturing is still a relatively new process and it lacks a heritage database from which to draw, highly conservative design allowables were used. Moreover, in addition to the usual safety factors employed for developing space products, an AM Conservatism Factor of 1.5 has also been incorporated, which allows for a 'comfort zone' to compensate for its lack of heritage data.

The Margin of Safety (MoS) of the part is calculated according to the formula below:

$$MoS_{yield} = \frac{\sigma_{allw,AlSi10}}{\sigma_{vonMises} \cdot SF_{yield} \cdot SF_{AM}} - 1 \tag{1}$$

where $\sigma_{allw,AlSi10Mg}$ is the maximum allowable yield stress (design allowable), $\sigma_{vonMises}$ is the Von Mises stress obtained from FEM analysis, SF_{yield} is the yield strength safety factor, and SF_{AM} is the

Additive Manufacturing conservatism safety factor. Equation (1) is used in subsequent sections to derive the Margin of Safety for each of the case studies presented.

3.4. Additive Manufacturing

All of the components in this paper were fabricated on an EOS M290 machine housed at Morf3D in El Segundo, California, which has a maximum power output of 370 W, and were built with a 30-micron layer thickness. The build plate was elevated to 165C in an effort to mitigate internal residual stresses and to eliminate the need for subsequent heat treatment, which reduces the mechanical values considerably. The material used was virgin AlSi10Mg powder characterized with D10, D50, and D90 values of 22.7, 41.8, and 69.8 microns, respectively. For clarity, D10 is the diameter at which 10% of the powder's mass is comprised of particles with a diameter less than this value, and so on for D50 and D90. The additive manufacturing environment was flushed with argon gas. Materialise Magics software was used to prepare the data and to generate support structures. EOSPRINT software was used to design and optimize build parameters.

Five Vertical and five horizontal tensile coupons, three density cubes, and three thin walled hermetically sealed powder archival components were added to each build plate along with the desired artifacts.

3.5. Mechanical and Material Verification

The next step in the holistic process flow is Testing. None of the components in this work required subsequent heat treatment due to the fact that they were built with an elevated build plate temperature of 165C. Hence the coupons were cut from the build plate with a wire EDM, and were sent directly to the testing laboratory. The co-fabricated coupons are tested for Ultimate Tensile Strength, Yield Strength, Elongation, and density, and their values must be greater than or equal to the acceptance allowables imposed by the customer.

After removing the components from the build plate with the wire EDM, their support structures were removed, and all accessible surfaces were media blasted. Further, attachment points and mating surfaces were machined to achieve the tolerances required for their function. The components or component assembly (e.g., case study 3) were subjected to vibration testing consisting of a combination of the following: low level sine, sine dwell, and random vibration.

In addition to structurally testing the components, each component was scanned via Computed Tomography (CT) in order to identify the internal pores, inclusion, or cracks within the resolution of the CT scanner.

Additionally, a 3D model of the fabricated components was generated with the CT scan data and compared to the nominal CAD file from which it was fabricated. A comparison of the two files was used to generate a heat map illustrating any geometric deviations as compared to the allowables set forth from the customer.

4. Case Studies

Three case studies are discussed that offer different insights into topology optimization. The heritage geometry of the first case study is illustrated in Figure 3, which is a Star Tracker Camera Bracket employed in SSTL's Technology's mission. The heritage design was machined out of titanium bulk. The constraints for this case study were that the attachment points remain fixed and that the design space could not extend further than the limits that were illustrated by the geometrical faces, however the bounding planar faces need not be preserved.

Figure 4 illustrates the second case study that is an edge insert also used in the SSTL Technology mission. It is apparent that the pictured geometry could be fabricated with subtractive means. Hence, in order to make this a suitable candidate part for AM, this case study sought to significantly lightweight the component as compared to its heritage counterpart. For this case, the boundary conditions specified are that the orthogonal faces are required to be preserved as they act as mating surfaces to the satellite sandwich panels. The material internal to the faces, however, is free to be redistributed through topology optimization. For the purpose of redistributing the material internal to the geometric faces, two critical design issues were needed to be addressed: (a) eliminate the need for internal support structures, since they would not be removable in the enclosed space and (b) create a way to remove the unprocessed powder.

Figure 3. Nominal star tracker camera bracket.

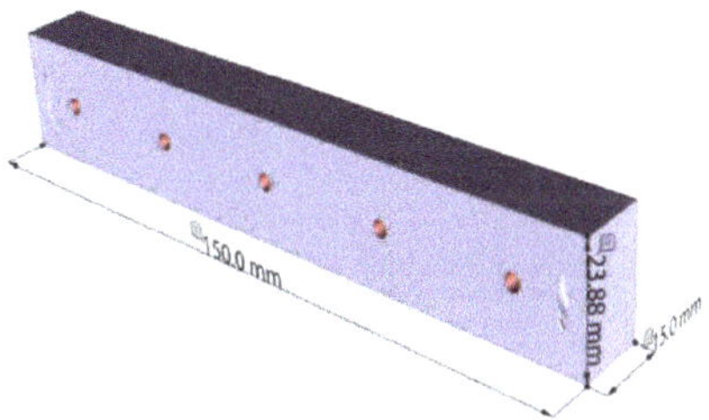

Figure 4. Nominal insert design geometry.

Figure 5 illustrates the nominal geometry of the third case study, which is a system of five parts that are assembled as one part. It serves as a thruster mount for a lunar vehicle intended for launch in the near future as part of SpaceIL's lunar mission. Once assembled it spans 800 mm in diameter and approximately 280 mm in height. It was chosen as a demonstration of topology optimization of an assembly of parts.

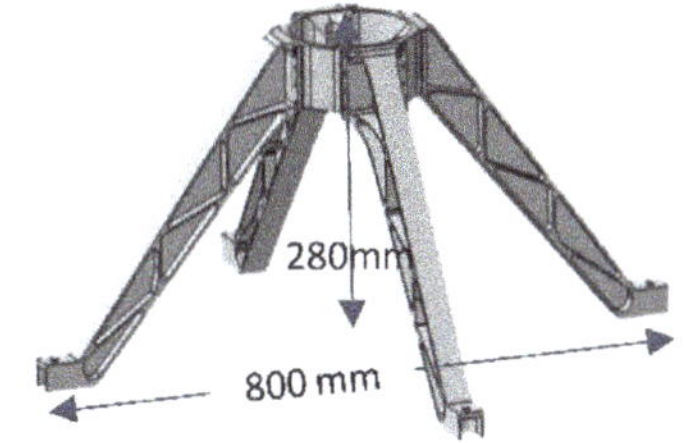

Figure 5. Nominal Lunar Lander engine mount.

The objective set for our optimization routine in each case is to minimize mass while remaining simultaneously compliant to stiffness requirements and maximum applied stress levels that are appropriate to each application. The case study examples are illustrative of the advantages of the union of topology optimization with Additive Manufacturing.

5. Results

The design, utilizing topology optimization (with the manual incorporation of the AM manufacturing considerations); FEM analysis, manufacture; and, testing results for the three case studies pictured in Figures 3–5 are discussed in detail below in the context of the holistic process flow. Each case study was selected for discussion in this work because they offer different topology optimization constraints.

5.1. Case Study 1: Star Tracker Camera Bracket

Case Study 1 presents the topology optimization example in which the material is free to move within the design space, and the boundaries of the design space (with the exception of the attachment points) need not coincide with the boundaries of the final part.

5.1.1. Topology Optimization

The design space for the Star Tracker Camera Bracket is illustrated in Figure 6. The geometry for the fixed interface points were taken from the CAD file. The design space was meshed with a mesh element size of 2 mm, and subsequently with 0.75 mm for refined analysis. The OptiStruct optimization routine was employed, with the material being given the freedom to be located anywhere within the design space. With the exception of the attachment points, the external boundaries of the design space are not required to be incorporated into the optimized part.

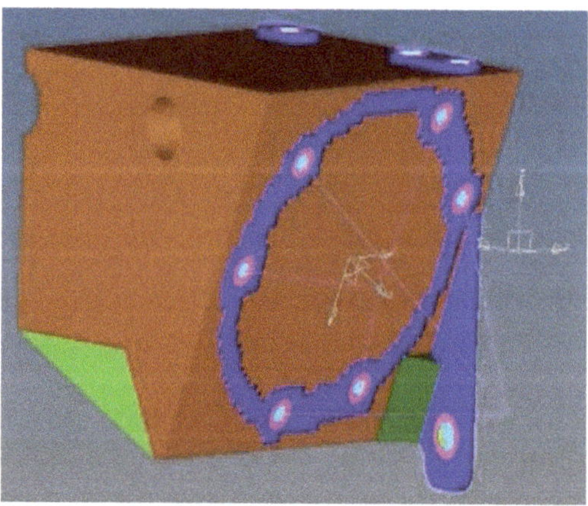

Figure 6. Design space for the Star Camera Bracket Topology Optimization Routine.

A minimum and maximum member thickness as 6 and 20 mm, respectively, was implemented. Other design constraints imposed that were within the topology optimization routine were a maximum stress of 125 MPa, and frequency constraint that the first eigenfrequency be greater than 140 Hz with a factor of safety of 1.3. The build orientation was determined by the engineer, and, once established, the other Additive Manufacturing constraints were considered by the engineer through manual iterations. To be self-supporting, care was taken to insure that no component grew at an angle less than 45 degrees with respect to the build plate, and efforts to minimize large volumetric jumps that would otherwise cause thermal jumps and subsequent component warping were also incorporated. Intermediate design results were interpreted and Poly-NURBs based software was employed to create a printable design.

5.1.2. FEM Design Verification

The final optimized design of the Tracker Camera bracket was subjected to a Finite Element Modeling campaign, as described in Figure 2, which serves as a redundant check that the final design will perform according to the loads prescribed. This constitutes the first verification cycle of the holistic process flow. If the topology optimized design fails the Finite Element Analysis, the design is modified by adding or redistributing mass and the design is analyzed again until it passes the FEM Design verification step. This is a necessary component in the holistic process flow because manual steps are incorporated into the topology optimization routine to account for the minimization of support structures, tool access, and other practical issues that are associated with manufacturing and assembly.

For the Star Tracker Camera bracket, RBE2 elements were used for each attachment point. Accelerations were implemented for each global axis of the satellite with $120g_{peak}$. Eigen frequencies in the range of 1–100 were calculated with the design restriction that the first eigenfrequency be greater than 140 Hz.

Figure 7 illustrates the results of the FEA analysis for the Star Tracker Camera bracket. Except for one attachment point (indicated by the blue arrow), all von Mises stresses were found to be below 110 MPa, well within the factor of safety.

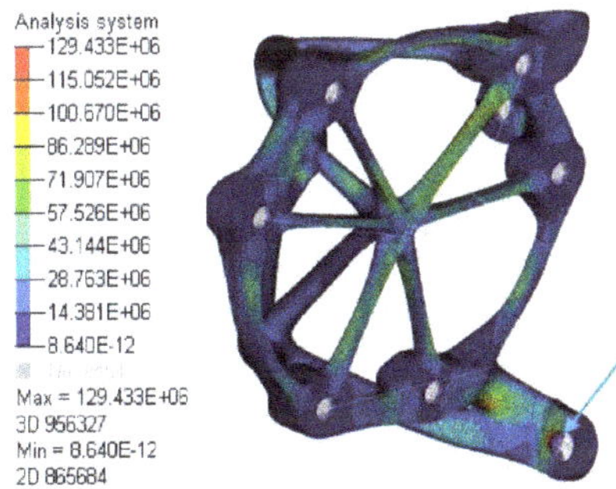

Figure 7. Results of FEM simulation analysis.

The maximum stress resulting from the model at the indicated attachment point is larger than the physical realization would be because the modeled bolt connection assumed that all nodes of the inner borehole were clamped along all six degrees of freedom. In the physical situation, however, a washer would clamp the part on one side of the attachment point and on the opposite side the part would be clamped to the spacecraft structure. Hence, a more realistic situation would require the modeling of the washer and implementing contact definitions, yielding a lower calculated stress. Because the focus of this project was not focused on FEM modeling of the component connections, but it is used as a guide to indicate the soundness of our topology optimization results, we chose to approximate the modeling of the connection point, as described. The soundness of the fabricated parts is subsequently validated through proof testing as described in further studies. Furthermore, the assumed material yield limit that is employed in the analysis of 190 MPa is also conservative, as testing of the manufactured witness coupons resulted in average values of 244.9 MPa and 208.5 MPa for horizontally and vertically grown coupons, respectively (see tensile testing below, Table 1), adding another layer of safety.

Table 1. Results from tensile testing campaign for Case Studies 1 and 2.

Build Orientation		Ultimate Strength (MPa)	Yield Strength (MPa)	% Elongation
Horizontal	Average	392.98	244.93	6.60
	Std. Dev.	8.30	7.85	0.55
Vertical	Average	394.29	208.54	5.5
	Std. Dev.	1.63	2.24	0.58

Applying Equation (1), with $\sigma_{vonMises}$ equal to 110 MPa, SF_{yield} equal to 1.1, $SF_{AM} = 1.5$, and $\sigma_{allw, AlSi10Mg} = 190$ MPa, yields a MoS of 0.05. It should be reiterated, however, that this Margin of Safety includes a somewhat arbitrary Safety Factor for Additive Manufacturing equal to 1.5 that has been added in this study to alleviate fears for this new technology. It is felt that this additional safety factor will evolve to values that are closer to 1.0 in the future.

The topologically optimized design is pictured in Figure 8. It can be seen that the optimized design represents a radical departure from the original bracket design, as the only fixed geometrical design constraints pertain to the attachment points between the bracket and the satellite panel and those between the bracket and the camera.

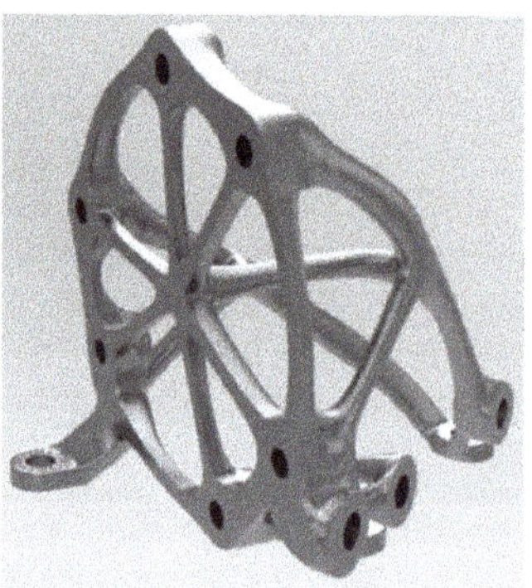

Figure 8. Final Optimization design.

The camera bracket design that is rendered in Figure 8 illustrates the benefits of Additive Manufacturing coupled with topology optimization. Its weight is 89 g as compared to 425 g for the heritage part, and the design took approximately two weeks to complete its design. The heritage part was manufactured out of Titanium. Based on the analysis carried out in this work, the heritage part appears to be over-engineered, as the loading and boundary conditions can be satisfied with an aluminum alloy component. Note that no finite element analysis was conducted on the heritage part.

The authors are hard-pressed to envision the design of the bracket pictured without the tools implemented in this work, and furthermore, even if the design were conceived, it would be extremely difficult (if even possible) and costly to manufacture with traditional subtractive methods.

5.1.3. Manufacturing

Figure 9 illustrates the build plate as is sits in the EOS M290 Additive Manufacturing platform after fabrication and after powder removal. It should be noted that the support structures are minimal. Visible are five vertical and five horizontal tensile specimens that were fabricated with the components, pyramid thin-walled powder archival vessels, and density cubes for evaluation. Figure 10 illustrates the as-built camera star bracket prior to machining of mating surfaces and attachment points. The circular hole pattern, as well as the other attachment points are to be machined in order to achieve their required tolerances.

5.1.4. Testing

The next step in the holistic process flow is testing. The tensile coupons were tested for Ultimate Tensile Strength, Yield Strength, Elongation according to ASTM standard E8, and the density cubes were evaluated via ASTM standard E 562 for microstructure and density. The results are illustrated in Table 1 and are well within the design allowables. Table 1 represents data from two build plates with 10 vertical and 10 horizontal tensile coupons, and six density cubes.

Figure 9. Build plate with 4star tracker camera bracket components.

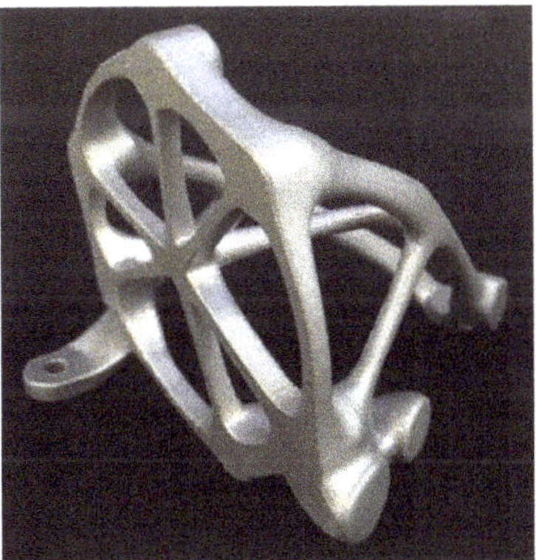

Figure 10. AM fabricated camera star bracket.

On a component level, the components were subjected to the testing campaign sketched in Figure 11. All results were exceptional.

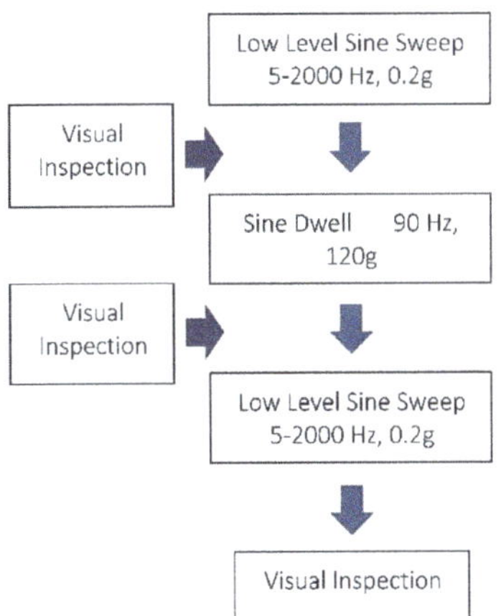

Figure 11. Structural testing flow.

In addition to structurally testing the components, they were each CT scanned and no internal pores, inclusion, or cracks were identified within the applied resolution of the CT scanner. The components were scanned with a resolution of 120 microns, hence it is important to note that defects that are smaller than 120 microns may exist but were undetectable with the CT modality. The CT scan was incorporated into the process flow as a risk-mitigation and process control check.

A 3D model was created from the Computed Tomography data of the printed part and compared to the CAD file from which it was printed. Overlapping the two files allows for the generation of heat maps in order to obtain a visual representation of the deviations of the fabricated component from the ideal geometry. The heat map is illustrated in Figure 12. It was found that the printed artifacts exhibited a tendency for small deviations on the downward facing edges with respect to the build direction, which can be attributed to the Additive Manufacturing process. This is because the downward facing edges are in direct contact with powder as opposed to solid, and powder is a less efficient thermal conductor than solid, resulting in significantly different heat transfer attributes between the down facing surfaces and the up facing surfaces. The maximum deviations between the Additively Manufactured part and the CAD data was measured to be under 280 microns. It is also noteworthy that the components were not required to be heat treated, as the measured deviations were within the design allowables specified by the customer. It is possible that a portion of the measured deviations are due to internal residual stresses in combination with the rougher downskin surfaces inherent to powder bed laser fusion. Hence, heat treatment may reduce the geometric deviations, however heat treatment will also be associated with reductions in ultimate and yield strength.

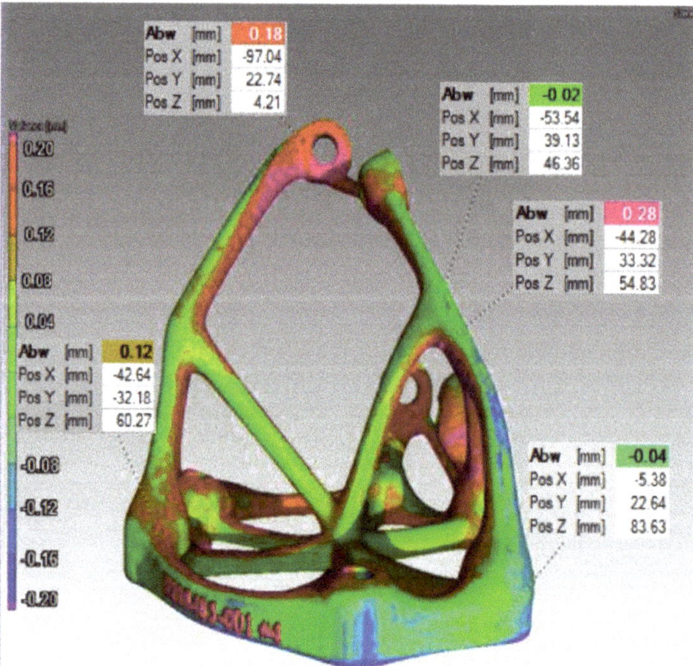

Figure 12. Overlay of nominal CAD geometry with fabricated part geometry obtained from CT scan data.

5.1.5. Case 1 Summary

The heritage and traditionally fabricated star tracker bracket illustrated in Figure 3 weighed 425 g and was machined out of Titanium. The topologically optimized component weighed 89 g and it was Additively Manufactured out of ALSi10Mg. Both were designed to withstand the same loading conditions and satisfy the same boundary conditions. Furthermore, the entire process, including:

topology optimization, fabrication, coupon, and component testing took only eight weeks, illustrating that significant lightweighting can be achieved on heritage parts in a rapid timeframe.

5.2. Case Study 2: SSTL Edge Insert

Case Study 2 provides the example in which the external boundaries must remain fixed, however material is free to move within the space in order to support the loading conditions.

5.2.1. Topology Optimization

Because the edge inserts are press fit into an opening of a satellite sandwich panel (with overall panel dimensions of 250 × 70 mm), the external boundaries of the inserts must remain fixed, as illustrated in Figure 4.

A 3 mm distance around each attachment point was considered to be fixed and 'non-design space', as well as 3 mm thickness for all of the orthogonal faces. Because it is anticipated that material will be displaced in the internal space within the external walls, methods to remove unmelted powder were incorporated into the design. Additionally, a build direction was selected so that proper structures can be incorporated to support horizontally built surfaces. The choice of build direction that is indicated in Figure 13 requires that the long horizontal surface be internally supported.

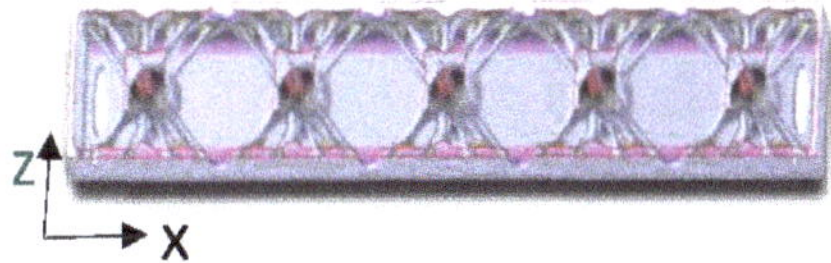

Figure 13. Schematic the edge insert cross section illustrating the macro-lattice supporting the top horizontal surface.

Several different design possibilities were explored and are discussed elsewhere [22]. The final design that is pictured in Figure 13 reveals an internal macro-lattice structure that is designed to support the long horizontal surface. The branch like structures are five across and three deep. Also incorporated are two holes on both ends of the insert for powder removal and serve to equilibrate the internal and external pressure of the structure; a necessity that is required for hardware entering the rarefied environment of space. Other than the supportive branches shown, the insert is hollow, leading to a significant decrease in weight.

5.2.2. FEM Design Verification

The sandwich panel in which the insert fits was modeled with two-dimensional (2D) shell elements for the face sheets and 3D elements for the core. The insert is connected to the face sheet and to the core via coincident nodes. The boundary conditions are that all cut faces of the sandwich panel (except for the top and bottom face and face where the insert is embedded) are simply supported (constraining DOF123). The edge inserts were only subject to the following static load cases on each of the five bolts: a force in the x direction (in-plane) of 225N and in out-of-plane (Z) direction a force of 355N.

Figure 14 illustrates the FEA results for the edge attachment. The maximum von Mises stress of 72.7 MPa was identified at the opening port (shown in red) that allows for pressure venting and powder removal, and it is due to the geometrical arrangement. The maximum von Mises stress in the branch like structure was less than 60 MPa, both of which are well below the design allowables. Applying Equation (1), with $\sigma_{vonMises}$ equal to 72.7 MPa, SF_{yield} equal to 1.1, $SF_{AM} = 1.5$, and $\sigma_{allw, \, AlSi10Mg} = 190$ MPa, yields a MoS of 0.58. It is important to note that the calculated margin or safety also includes an AM safety factor of 1.5 to satisfy industrial concerns.

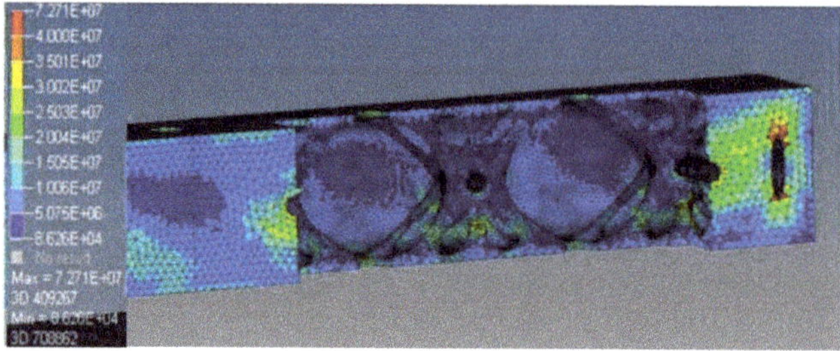

Figure 14. FEA results of Edge Insert.

5.2.3. Manufacturing and Testing

The inserts and camera star bracket were manufactured together with the same in-process test coupons. Hence, the mechanical results that are provided above in Table 1 also apply to Case Study 2. The edge inserts were not tested as a part of this project as they were tested on a system level by the customer. Unfortunately, the authors do not have access to the satellite system testing data. The tensile results in Table 1 were a product of 10 vertical and 10 horizontal tensile coupons over two builds. The final sectioned component is pictured in Figure 15.

Figure 15. As-printed image of the sectioned Edge insert.

5.2.4. Case 2 Summary

The topologically optimized edge insert experienced a 40% reduction in weight over the heritage solid counterpart. Its design was guided in part by its build orientation, the need to be self-supporting, and the necessity of removing unsintered powder.

5.3. Case Study 3: Lunar Lander Engine Mount

Case Study 3, as pictured in Figure 5, provides an illustrative example of topology optimization of a system of large components.

5.3.1. Topology Optimization

Preliminary explorations of the topology optimization exercise indicated that in order to satisfy the loading and stiffness requirements, the solid legs of the original nominal design (Figure 5) tended to split into three separate branches, each of which attached to the central hub. It was also discovered that the optimization routine sought solutions in which the branch type structures attached to the central hub at two elevations, arranged as far as possible from each other. Hence, it was decided to explore this avenue of topology optimization and to study whether this solution was mechanically viable. These results also enabled a reduction of the design volume in further optimization runs for the sake of computational effort and for the assessment of the viability of certain design solutions. The design space (gray shaded region) for the topological optimization exercise of the lunar launcher engine mount and the preliminary optimization result that fulfills the strength and stiffness requirements is illustrated in Figures 16 and 17. Figure 16 is the top view illustrating how each leg is now formed as three branches and Figure 17 is the isometric image illustrating how the legs attach to the central hub. The apparent discontinuities in the solution are due to the density sensitivity chosen for the image.

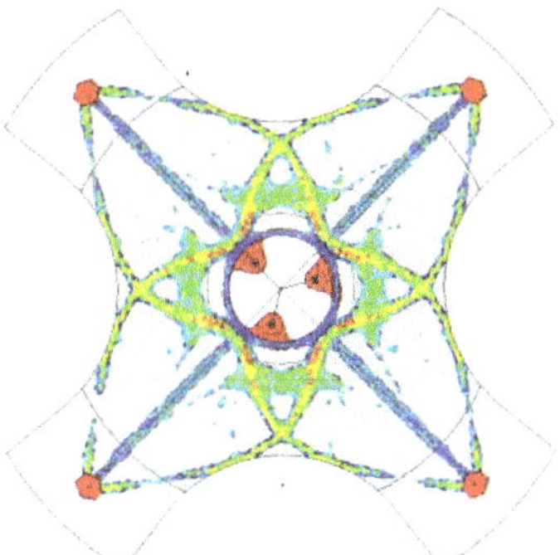

Figure 16. Top design space (gray shaded region) and initial topology optimization result.

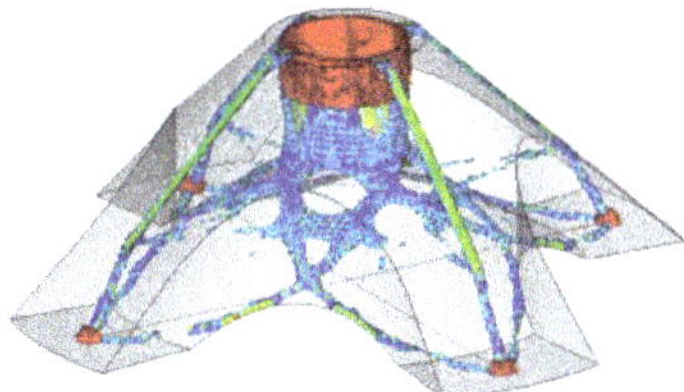

Figure 17. Iso view of design space (gray shaded region) and initial topology optimization result.

These initial studies were focused around the topology optimization of one integrated component that includes the connected legs and hub, with the understanding that once the topology optimization design concept was established, it would be determined where and how to split the components in order to fabricate them. Since the size of the components was a design driving factor, a final topology optimization was performed that incorporated the determined connection points for the legs and the hub and forces the topology optimization to pass through these connections.

An important aspect of the design was to create self-supporting components, or when not possible, components with the minimal number of support structures. Hence, the design was driven in part by component orientation with respect to build direction to ensure minimal support structures. Figure 18 illustrates a rendering of the final design of five split components that are bolted together by close tolerance shear bolts. Each of the components have been designed such that they fit within the build chamber of the EOS M290 machine. Moreover, the legs were designed so that two of them fit together within the dimensions of the build chamber, reducing the number of print jobs to three, consisting of: two plates with two legs each and one with the hub. Additionally, design efforts that are focused on creating approximately uniform member cross sections in an effort to achieve nearly homogeneous heat transfer towards the base plate, thereby minimizing the occurrence of thermal related warping. Furthermore, since downward facing surfaces are known to exhibit inferior characteristics in terms of surface quality and hence dynamic performance, these regions not chosen for highly loaded areas.

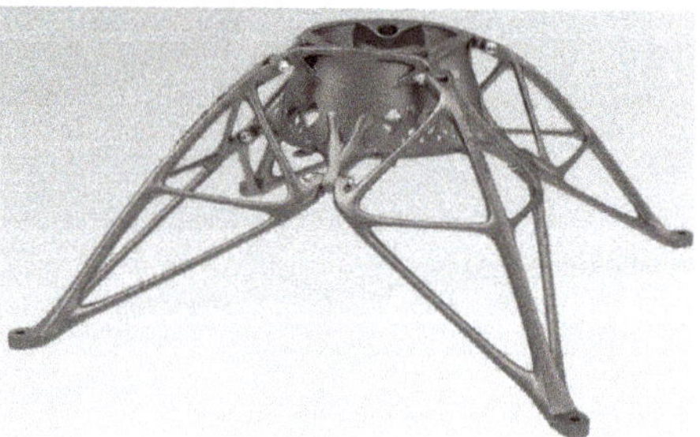

Figure 18. Rendering of final topologically optimized design.

5.3.2. FEM Design Verification

Similar to Case Studies 1 and 2, the AM parts are modeled with solid elements (tetra). Integrating components (not shown) that are not additively manufactured are an expansion cone and heat shield, which are modeled using shell elements. The connection of the components was modeled through close tolerance shear bolt connections. More details about the integrating components can be found elsewhere [4]. The engine FE model was created in detail to achieve a modal performance that matches with modal results that were provided by the engine supplier. The integrating heatshield is modeled with larger elements, as it is only used as mass representation.

The engine mounting structure is clamped at all four feet (at the inner bore hole), constraining all translational and rotational degrees of freedom to reflect the mounting to the spacecraft structure. Results from the dynamic testing campaign will be fed back to this step and the settings will be adjusted if necessary in order to correlate the model. The components are connected through close tolerance shear bolt connections. More details of how connections between the structural components were realized in the FE model are provided elsewhere [4].

The structure is subjected to quasi static as well as sine and random vibration loads. Results of the analysis are presented in Figures 19 and 20, which illustrate the stress plots for static equivalent load cases based on the accelerations from the random analysis ($3 \times g_{RMS}$ method) in the xy plane and z planes of the spacecraft, respectively. Enveloping the two load cases, the maximum stress was found to be 111.2 MPa. Using values of SF_{yield} equal to 1.1, SF_{AM} = 1.5, and $\sigma_{allw, AlSi10Mg}$ = 190 MPa, a MoS of 0.04 is calculated. Hence, all stresses are well within the design allowables for the final optimized concept providing positive margin of safety, and hence according to the holistic process flow, this design is analytically verified and ready to be manufactured. The static equivalent load level for the z-excitation is 154 g out of plane and 25 g in plane and 36 g out of plane and 79 g in plane for the xy-excitation, respectively. Moreover, fatigue and shock analysis were also performed. Fatigue analysis has been done with the ESA software ESAFATIG v4.3.1a; the fatigue calculations are based on the linear damage accumulation (Palmgren-Miner) rule. The fatigue study required that four times the life needs to be shown, and the analysis resulted in 10.4 times life. The shock assessment was based on the ECSS "Point Source Excitation Method". With a safety factor of 3 dB, a positive margin of 4.3 dB was determined from the analysis.

5.3.3. Manufacturing

As with Case Studies 1 and 2, all components were manufactured in AlSi10Mg on an EOS M290 system at the Morf3D facility. They were fabricated with a 30 micron [1.2 mils] build layer thickness and with a build plate temperature of 165 °C [329°F]. None of the components required heat treatment due to the design and due to the elevated build temperature. The powder that was used is characterized by the following D values: D50 of 42.78 microns [~1.7 mils] and D90 of 69.88 microns [~2.75 mils].

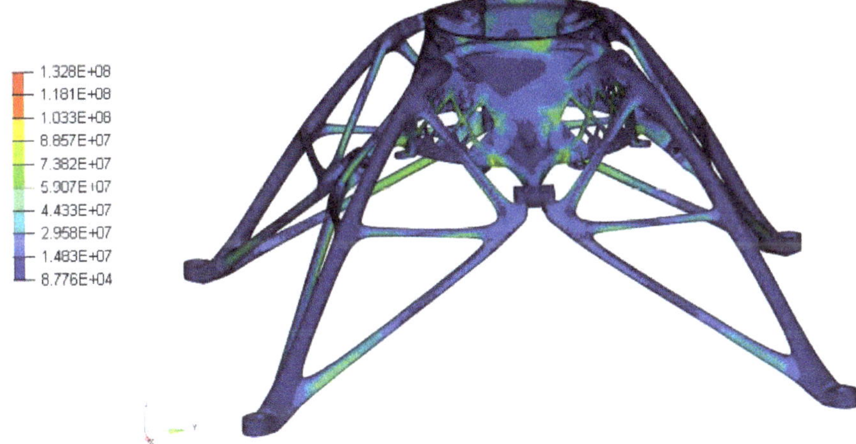

Figure 19. Results of the FEM analysis: Stress plot of the complete engine mount structure subjected to x/y excitation.

Figure 20. Results of the FEM analysis: Stress plot of the complete engine mount structure subjected to Z excitation.

Figure 21 depicts a photograph of the three build plates that comprise the system. The leftmost plate contains two legs, the center contains the connecting hub, and the rightmost contains the remaining two legs. In-process coupons (tensile and density cubes) were built with each plate along with the aforementioned pyramid powder archival system. The vertical coupons built with the hub were incorporated under the eight tabs around the main circumference that would otherwise require support structures, hence these coupons serve two purposes.

As can be seen in Figure 21, the design succeeds in terms of requiring minimal support structures to support extended overhangs, and it was possible to fabricate two legs simultaneously. After complete powder removal, the parts are removed from the plate with an EDM wire cutter. Hence, supports connecting the parts to the plate must have sufficient height to permit this subtractive removal process from the build plate.

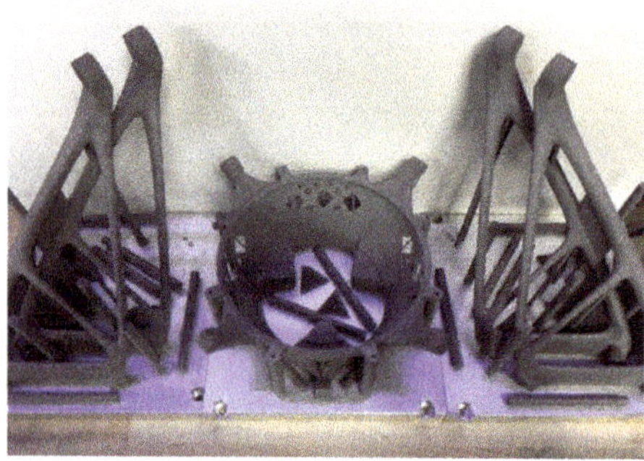

Figure 21. Photograph of three build plates after additive manufacturing.

Post processing included media blasting with glass bead media of 60–120 grit, followed by 70–140 grit. The attachment points of all components were machined to precision mating surfaces. Hence, in addition to designing and fabricating the desired components, it is necessary to design and fabricate jigs and tools for subsequent subtractive machining and testing.

5.3.4. Testing

Tensile tests for Case Study 3 are presented in Table 2, and they are comprised of 15 horizontal and 15 vertical coupons. Tested were the Ultimate Tensile Strength, Yield Strength, and Elongation for horizontally and vertically printed tensile coupons. All tensile testing was performed according to ASTM E8 standards. Coupons were fabricated as solid cylinders, as pictured in Figure 20 and machined to the appropriate dog-bone geometry required for testing. All tensile tests were well within the design allowables specified by the customer.

Table 2. Results from tensile testing campaign for Case Study 3.

Build Orientation		Ultimate Strength (MPa)	Yield Strength (MPa)	% Elongation
Horizontal	Average	396.2	259.6	8.1
	Std. Dev.	7.3	7.4	0.8
Vertical	Average	434.1	244.9	5.8
	Std. Dev.	11.8	9.0	0.7

Additionally, density coupons were evaluated via ASTM standard E 562 and it was found that for each case that the density was greater than 99.7%. Hence, the mechanical and material properties are acceptable and the testing continued on the component level.

Component testing included non-destructive CT scanning, which indicated that no pores, crack, or inclusions were identified within the applied resolution of the CT scanner, which was 150 microns. Hence, pores, cracks, and other flaws smaller than 150 microns may be present in the components. Additionally, as with the previous two test cases, the CT scanned data also yielded 3D models of the components that were compared to the nominal CAD files. Comparison of the actual to nominal hub geometry is illustrated in Figure 22, and that of one representative leg is found in Figure 23. It was found that all components were well within the design tolerances that were established by the customer.

Dynamic structural testing was performed on the assembled system and it included a low level sine sweep on each spacecraft axes as well as high level sine, random, and shock testing. The additively manufactured system of components performed well and they were determined to be structurally sound and suitable for flight.

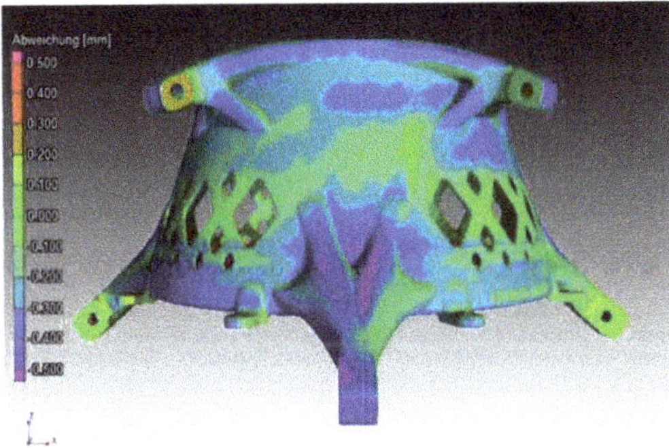

Figure 22. Actual versus nominal comparison of the hub from the CT data.

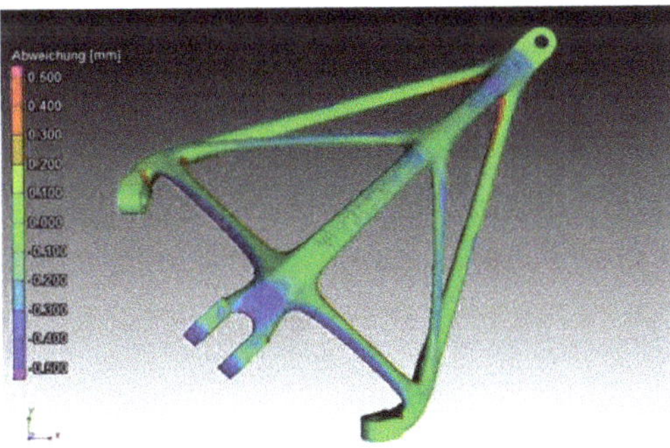

Figure 23. Actual versus nominal comparison of a leg from the CT data.

5.3.5. Case 3 Summary

Specific attention has been paid on designing for AM, including topology optimization and design interpretation for AM. Optimization goals included the minimization of component mass while meeting constraints for the first natural frequency as well as the maximum stress limit. Coupled design interpretation considerations addressed the segmentation of the large part into a system of components that fit within the build chamber, and the minimization of support structures.

The resulting system of components include four bionic inspired legs and a central hub that are significantly lighter in weight than the original baseline design (i.e., 2.95 kg compared to 4.0 kg). Note that the nominal un-optimized component and the topologically optimized component are both designated to be fabricated in AlSi10Mg. The weight savings in the topologically optimized component that are shown in Figure 18 is due to the optimization of material placement. In-process test coupons were fabricated with each build and their tested values were well within the design acceptance limits set by the customer. Powder archival units were also fabricated and stored for future interrogation if required. Computed Tomography testing on the component level revealed that the fabricated components were within the geometric design limits and there was negligible porosity, corroborating the density coupon testing at 99.7%.

6. Summary

Three case studies were selected to demonstrate the AM holistic process flow. The parts were selected because they qualified as good candidate parts. That is, they found value in reducing the component weight; they demonstrated a rapid fabrication life-cycle (including analysis, fabrication, testing, and certification); their required tolerances were such that they did not require heat treatment.

The first and possibly most important step in the process flow is the identification of suitable parts for additive manufacturing. Parts that are critically loaded may require a larger margin of safety, and those requiring tighter tolerances may require heat treatment to mitigate warping, or more extensive post-machining processes. Such parts may enjoy less weight savings than those that are presented in this work.

The aim of this work was to demonstrate the value of topology optimization combined with Additive Manufacturing for the fabrication of flight hardware with the aid of three unique case studies. Each case study offered a different example of topology optimization. The first case study permitted significant material re-location within the design space and did not require the incorporation of the exterior boundaries into the design, thereby resulting in an organically shaped component that bears little resemblance to its heritage counterpart and is significantly lighter in weight. The second case study required the preservation of the external boundaries of the design space into the actual design, but it allowed the topology optimization routine to remove a significant amount of material internal to the boundaries. Hence, externally this part resembled its heritage part, however its internal features were not possible to achieve with traditional manufacturing methods. The third case study illustrated a system of five components that worked together to create a lunar lander engine mount. In each case, care was taken to minimize support structures, and when supports were necessary, to ensure that they were accessible for removal. Additionally, care was taken to ensure that downward facing surfaces were not load bearing and for case study 2, any unmelted powder was readily removable from the component.

It was found in each case that significant weight savings occurred in the topologically optimized version. It should be noted that the particular heritage components that were selected for the presented case studies were susceptible to weight savings through redesign (e.g., the component presented in the first case study need not be fabricated out of Titanium when Aluminum would support the intended application). Future weight savings may not be as significant as the examples presented in this work.

In addition to coupling the design with the build orientation, downstream processes, such as testing and secondary machining, were considered in the design phase so that methods to fixture the resulting geometries in their appropriate jigs were accommodated. The three case studies also demonstrated the efficacy of a holistic process-flow that originates with part selection and includes topology optimization, FEM validation, Additive Manufacturing, mechanical and material testing, and terminates in approving the component for flight. The process flow was successfully demonstrated with three case studies; each of which with different topology optimization requirements. The entire process flow for the first two case studies, from concept to certification, took only eight weeks to complete. The third case study took 16 weeks from concept to certification. The aggressive testing campaign included tensile tests, density testing, CT scanning, and dynamic testing. The Additively Manufactured components passed all testing criteria with comfortable margins, demonstrating that additive manufacturing, when approached with a methodical holistic process flow, is an attractive manufacturing method for creating lightweight, functional metallic components that can be approved or flight.

Author Contributions: Conceptualization, M.F., M.G; I.M., M.O., methodology, M.F., M.G., Resources, I.M., M.G., formal analysis, M.F., investigation, M.O., data curation, M.O., M.G., writing, M.O. project administration, M.O., I.M., M.G.

Funding: This research received no external funding.

Acknowledgments: The author would like to acknowledge Russel Vernon of Altair Engineering for his collaborative efforts on this project regarding topology optimization of Case Study 2. Special thanks are also extended to the executive management team at Electro Optical Systems, EOS GmbH, for their continued support of this work and to Morf3D for their continued support.

Conflicts of Interest: The authors declare o conflicts of interest.

References

1. Wohlers, T. *Additive Manufacturing and 3D Printing State of the Industry: Wohlers Report*; Wohler's Associates, Inc.: Fort Collins, CO, USA, 2016.
2. Zhai, Y.; Lados, D.; Lagoy, J. Additive Manufacturing: Making Imagination the Major Limitation. *JOM* **2014**, *6*, 808–816. [CrossRef]
3. Orme, M.; Gschweitl, M.; Ferarri, M.; Vernon, R.; Madera, I.; Yancey, R.; Mouriaux, F. Additive Manufacturing of Lightweight, Optimized, Metallic Components Suitable for Space Flight. *AIAA J. Spacecr. Rocket.* **2017**, *54*, 1050–1059. [CrossRef]
4. Orme, M.; Gschweitl, M.; Ferrari, M.; Madera, I.; Mouriaux, F. Designing for Additive Manufacturing: Lightweighting Through Topology Optimization Enables Lunar Spacecraft. *J. Mech. Des.* **2017**. [CrossRef]
5. Deckard, C.; Beaman, J.; Darrah, J.F. Method for Selective Laser Sintering with Layerwise Cross-Scanning. U.S. Patent US 5155324 A, 13 October 1986.
6. Horn, T.; Harrysson, O. Overview of current additive manufacturing technologies and selected applications. *Sci. Rev.* **2012**, *95*, 255–282. [CrossRef]
7. Agarwala, M.; Bourell, D.; Beaman, J.; Marcus, H.; Barlow, J. Direct selective laser sintering of metals. *Rapid Prototyp. J.* **1995**, *1*, 26–36. [CrossRef]
8. Das, S. Physical Aspects of Process Control in Selective Laser Sintering of Metals. *Adv. Eng. Mater.* **2003**, *5*, 701–711. [CrossRef]
9. Slotwinski, J.A.; Garboczi, E.J.; Stutzman, P.E.; Ferraris, C.F.; Watson, S.S.; Peltz, A.M.A. Characterization of Metal Powders Used for Additive Manufacturing. *J. Res. Natl. Inst. Stand. Technol.* **2014**, *119*, 460–493. [CrossRef] [PubMed]
10. Frazier, W. Metal Additive Manufacturing: A review. *J. Mater. Eng. Perform.* **2014**, *23*, 1917–1928. [CrossRef]
11. Herzog, D.; Seyda, V.; Wycisk, E.; Emmelmann, C. Additive manufacturing of metals. *Acta Mater.* **2016**, *117*, 371–392. [CrossRef]
12. Louvis, E.; Fox, P.; Sutcliffe, C. Selective laser melting of aluminum components. *J. Mater. Process. Technol.* **2010**, *211*, 275–284. [CrossRef]
13. Olakanmi, E.; Cochrane, R.; Dalgarno, W. A review on selective laser sintering/melting (SLS/SLM) of aluminum alloy powders: Processing, microstructure, and properties. *Progr. Mater. Sci.* **2015**, *74*, 401–477. [CrossRef]
14. Sercomb, T.B.; Li, X. Selective laser melting of aluminum and aluminum metal matrix composites: Review. *Mater. Technol.* **2016**, *31*, 77–85.
15. Song, B.; Zhao, X.; Li, S.; Han, C.; Wei, Q.; Wen, S.; Liu, J.; Shi, Y. Differences in microstructure and properties between selective laser melting and traditional manufacturing for fabrication of metal parts; A review. *Front. Mech. Eng.* **2015**, *10*, 111–125. [CrossRef]
16. Trevisan, F.; Calignano, F.; Lorusso, M.; Pakkanen, J.; Aversa, A.; Ambrosio, E.P.; Lombardi, M.; Fino, P.; Manfredi, D. On the Selective Laser Melting (SLM) of the AlSi10Mg Alloy: Process, Microstructure, and Mechanical Properties. *Materials* **2017**, *10*, 76. [CrossRef] [PubMed]
17. Kempen, K.; Thijs, L.; Van Humbeeck, J.; Kruth, J.-P. Mechanical properties of ALSi10Mg produced by Selective Laser Melting. *Phys. Procedia* **2012**, *39*, 439–446. [CrossRef]
18. Prashanth, K.; Scudino, S.; Klauss, H.; Surreddi, K.; Lober, L.; Wang, Z.; Chaubey, A.; Kuhn, U.; Eckert, J. Microstructure and mechanical properties of Al-12Si produced by selective laser melting: Effect of heat treatment. *Mater. Sci. Eng. A* **2014**, *590*, 153–160. [CrossRef]
19. Vranken, B.; Thijs, L.; Kruth, J.-P.; Humbeeck, J.V. Heat treatment of Ti6Al4V produced by Selective Laser Melting; Microstructure and mechanical properties. *J. Alloys Compd.* **2012**, *541*, 117–185. [CrossRef]
20. Li, W.; Shuai, L.; Liu, J.; Zhang, A.; Zhou, Y.; Wei, W.; Yan, C.; Shi, Y. Effect of heat treatment on AlSi10Mg alloy fabricated by selective laser melting: Microstructure evolution, mechncial properties and fracture mechanism. *Mater. Sci. Eng. A* **2016**, *663*, 116–125. [CrossRef]
21. Aboulkhair, N.; Everitt, N.; Ashcroft, I.; Tuck, C. Reducing porosity in AlSi10Mg parts processed by selective laser melting. *Addit. Manuf.* **2014**, *1*, 77–86. [CrossRef]
22. Orme, M.; Gschweitl, M.; Vernon, R.; Ferrari, M.; Madera, I.; Yancey, R.; Mouriaux, F. A Demonstration of Additive Manufacturin as an Enabling Technology for Rapid Satellite Design and Fabrication. In Proceedings of the SAMPE, Long Beach, CA, USA, 23–26 May 2016.

23. Rozvany, G. *Topology Optimization in Structural Mechanics*; CISM International Centre for Mechanical Sciences; Springer: Berlin, Germany, 1997; p. 322.
24. Brennan, J. 20 Years of Topology Optimization: Birth and Maturation of a Disruptive Technology. 2015. Available online: http://insider.altairhyperworks.com/20-years-topology-optimization-birth-maturation-disruptive-technology/ (accessed on 27 January 2017).
25. Bracket, D.; Ashcroft, I.; Hague, R. Topology Optimization for Additive Manufacturing. In Proceedings of the Solid Freeform Fabrication Symposium, Austin, TX, USA, 17 August 2011.
26. Sigmund, O. On the usefulness of non-gradient approaches in topology optimization. *Struct. Multidiscip. Optim.* **2011**, *43*, 589–596. [CrossRef]
27. Yan, C.; Hao, L.; Hussein, A.; Bubb, S.L.; Young, P.; Raymont, D. Evaluation of light-weight AlSi10Mg periodic cellular lattice structures fabricaated via direct metal laser sintering. *J. Mater. Process. Technol.* **2014**, *214*, 856–864. [CrossRef]
28. Langelaar, M. Topology optimization of 3D self-supporting structures for additive manufacturing. *Addit. Manuf.* **2016**, *12*, 60–70. [CrossRef]
29. Mirzendehdel, A.M.; Krishnan, S. Support structure constrained topology optimization for additive manufacturing. *Comput.-Aided Des.* **2016**, *81*, 1–13. [CrossRef]
30. Allaire, G.; Dapogny, C.; Estevez, R.; Faure, A.; Michailidis, G. Structural optimization under overhang constraints imposed by additve manufacturing technologies. *J. Comput. Phys.* **2017**, *351*, 295–328. [CrossRef]
31. Guo, X.; Zhou, J.; Zhang, W.; Du, Z.; Liu, C.; Liu, Y. Self-supporting strucutre design in additive manufacturing through explicit topology optimiation. *Comput. Methods Appl. Mech. Eng.* **2017**, *323*, 27–63. [CrossRef]
32. Wang, Y.; Gao, J.; Kang, Z. Level set-based topology optimization with overhang constraing: Towards support-free additive manufacturing. *Comput. Methods Appl. Mech. Eng.* **2018**, *329*, 591–614. [CrossRef]
33. Mhapsekar, K.; McConaha, M.; Anand, S. Additive Manufacturing Constraints in Topology Optimization for Improved Manufacturability. *J. Manuf. Sci. Eng.* **2018**, *140*, 051017. [CrossRef]
34. Garaigordobil, A.; Ansola, R.; Santamaria, J.; Fernandez de Bustos, I. A new overhang constraint for topology optimization of self-supporting structures in additive manufacturing. *Struct. Multidicip. Optim.* **2018**, *58*, 2003–2017. [CrossRef]
35. Jankovics, D.; Gohari, H.; Tayefeh, M.; Barari, A. Developing Topology Optimization with Additive Manufacturing Constraints in ANSYS®. *IFAC Pap. OnLine* **2018**, *51*, 1359–1364. [CrossRef]
36. Mezzadri, F.; Bouriakov, V.; Qian, X. Topology optimization of self-supporting suport structures for additive manufacturin. *Addit. Manuf.* **2018**, *21*, 666–682. [CrossRef]
37. Liu, J.; Gaynor, A.; Chen, S.; Kang, Z.; Suresh, K.; Takezawa, A.; Li, L.; Kato, J.; Tang, J.; Want, C.; et al. Current and future rends in topology optimizatin for additive manufacturing. *Struct. Multidiscip. Optim.* **2018**, *57*, 2457–2483. [CrossRef]

MDPI

St. Alban-Anlage 66

4052 Basel

Switzerland

Tel. +41 61 683 77 34

Fax +41 61 302 89 18

www.mdpi.com

Designs Editorial Office

E-mail: designs@mdpi.com

www.mdpi.com/journal/designs